Lecture Notes
in Business Information Processing

67

Series Editors

Wil van der Aalst
Eindhoven Technical University, The Netherlands
John Mylopoulos
University of Trento, Italy
Michael Rosemann
Queensland University of Technology, Brisbane, Qld, Australia
Michael J. Shaw
University of Illinois, Urbana-Champaign, IL, USA
Clemens Szyperski
Microsoft Research, Redmond, WA, USA

Jan Mendling Matthias Weidlich
Mathias Weske (Eds.)

Business Process Modeling Notation

Second International Workshop, BPMN 2010
Potsdam, Germany, October 13-14, 2010
Proceedings

 Springer

Volume Editors

Jan Mendling
Humboldt-Universität zu Berlin
Institut für Wirtschaftsinformatik
Unter den Linden 6, 10099 Berlin, Germany
E-mail: contact@mendling.com

Matthias Weidlich
Mathias Weske
University of Potsdam
Hasso Plattner Institute for Software Systems Engineering
Prof.-Dr.-Helmert-Str. 2-3, 14482 Potsdam, Germany
E-mail: {matthias.weidlich,mathias.weske}@hpi.uni-potsdam.de

Library of Congress Control Number: 2010935672

ACM Computing Classification (1998): J.1, H.3.5, H.4, D.2

ISSN   1865-1348
ISBN-10   3-642-16297-5 Springer Berlin Heidelberg New York
ISBN-13   978-3-642-16297-8 Springer Berlin Heidelberg New York

springer.com

© Springer-Verlag Berlin Heidelberg 2010

Typesetting: Camera-ready by author, data conversion by Scientific Publishing Services, Chennai, India
Printed on acid-free paper   SPIN: 06/3180   5 4 3 2 1 0

Preface

The BPMN 2010 workshop series provides a forum for academics and practitioners that share an interest in business process modeling using Business Process Modeling Notation (BPMN) which has seen a huge uptake in both academia and industry. It is seen by many as the de facto standard for business process modeling. It has become very popular with business analysts, tool vendors, practitioners, and end users. BPMN promises to bridge business and IT, and brings process design and implementation closer together.

BPMN 2010 was the second workshop of the series. It took place October 13–14, 2010 at the Hasso Plattner Institute at the University of Potsdam, Germany. This volume contains six contributed research papers that were selected from 16 submissions. There was a thorough reviewing process, with each paper being reviewed by, on average, four Program Committee members. In addition to the contributed papers, these proceedings contain three short papers and three extended abstracts of the invited keynote talks. In conjunction with the scientific workshop, a practitioners' event took place the day after the workshop.

We want to express our gratitude to all those who made BPMN 2010 possible by generously and voluntarily sharing their knowledge, skills, and time. In particular, we thank the Program Committee members as well as the additional reviewers for devoting their expertise and time to ensure the high quality of the workshop's scientific program through an extensive review process. Finally, we are grateful to all the authors who showed their appreciation and support for the workshop by submitting their valuable work to it.

October 2010

Jan Mendling
Matthias Weidlich
Mathias Weske

Conference Organization

Program Chairs

Jan Mendling
Mathias Weske

Program Committee

Wil van der Aalst, The Netherlands
Gero Decker, Germany
Remco Dijkman, The Netherlands
Marlon Dumas, Estonia
Philip Effinger, Germany
Dirk Fahland, Germany
Jakob Freund, Germany
Denis Gagné, Canada
Félix García, Spain
Luciano García-Bañuelos, Estonia
Alexander Grosskopf, Germany
Thomas Hettel, Australia
Marta Indulska, Australia
Jana Koehler, Switzerland
Oliver Kopp, Germany
Agnes Koschmider, Germany
Frank Michael Kraft, Germany
Ralf Laue, Germany

Niels Lohmann, Germany
Bela Mutschler, Germany
Markus Nüttgens, Germany
Andreas Oberweis, Germany
Chun Ouyang, Australia
Karsten Ploesser, Australia
Frank Puhlmann, Germany
Jan Recker, Australia
Manfred Reichert, Germany
Hajo Reijers, The Netherlands
Stefanie Rinderle-Ma, Austria
Lucinéia Heloisa Thom, Brazil
Hagen Voelzer, Switzerland
Barbara Weber, Austria
Stephen White, USA
Karsten Wolf, Germany
Peter Wong, The Netherlands

Local Organization Committee

Matthias Weidlich
Katrin Heinrich

External Reviewers

Frank Hogrebe
Maria Leitner
Martina Peris
Christian Stahl

Table of Contents

Invited Talks

Unraveling Unstructured Process Models 1
  Marlon Dumas, Luciano García-Bañuelos, and Artem Polyvyanyy

BPEL vs. BPMN 2.0: Should You Care? 8
  Frank Leymann

An Overview of BPMN 2.0 and Its Potential Use 14
  Hagen Völzer

Full Papers

BPMN 2.0 Execution Semantics Formalized as Graph Rewrite Rules . . . 16
  Remco M. Dijkman and Pieter Van Gorp

On a Study of Layout Aesthetics for Business Process Models Using
BPMN 31
  Philip Effinger, Nicole Jogsch, and Sandra Seiz

The Role of BPMN in a Modeling Methodology for Dynamic Process
Solutions 46
  Jana Koehler

A Concept for Spreadsheet-Based Process Modeling 63
  Stefan Krumnow and Gero Decker

Managing Complex Event Processes with Business Process Modeling
Notation 78
  *Steffen Kunz, Tobias Fickinger, Johannes Prescher, and
  Klaus Spengler*

Managing Business Process Variants at eBay 91
  Emilian Pascalau and Clemens Rath

Short Papers

Managing Variability in Process Models by Structural
Decomposition 106
  Maria Rastrepkina

Adapting BPMN to Public Administration   114
    Victoria Torres, Pau Giner, Begoña Bonet, and Vicente Pelechano

An Evaluation of BPMN Modeling Tools   121
    Zhiqiang Yan, Hajo A. Reijers, and Remco M. Dijkman

Author Index   129

Unraveling Unstructured Process Models

Marlon Dumas[1], Luciano García-Bañuelos[1], and Artem Polyvyanyy[2]

[1] Institute of Computer Science, University of Tartu, Estonia
{marlon.dumas,luciano.garcia}@ut.ee
[2] Hasso Plattner Institute at the University of Potsdam, Germany
Artem.Polyvyanyy@hpi.uni-potsdam.de

Abstract. A BPMN model is well-structured if splits and joins are always paired into single-entry-single-exit blocks. Well-structuredness is often a desirable property as it promotes readability and makes models easier to analyze. However, many process models found in practice are not well-structured, and it is not always feasible or even desirable to restrict process modelers to produce only well-structured models. Also, not all processes can be captured as well-structured process models. An alternative to forcing modelers to produce well-structured models, is to automatically transform unstructured models into well-structured ones when needed and possible. This talk reviews existing results on automatic transformation of unstructured process models into structured ones.

1 Introduction

Although BPMN process models may have almost any topology, it is often preferable that they adhere to some structural rules. In this respect, a well-known property of process models is that of *well-structuredness*, meaning that for every node with multiple outgoing arcs (a *split*) there is a corresponding node with multiple incoming arcs (a *join*), such that the set of nodes between the split and the join form a single-entry-single-exit (SESE) region. For example, the process model shown in Fig.1(a) is unstructured because the parallel split gateways do not satisfy the above condition. Fig.1(b) shows an equivalent structured model.

The automatic transformation of unstructured process models into structured ones has been the subject of many R&D efforts. This keynote paper summarizes some of the results of these efforts, including the initial results on an ongoing research effort aiming at developing a complete method for structuring (BPMN) process models. But before discussing how to structure BPMN process models, let us briefly discuss why should we care about doing so.

2 Structured BPMN Models: Why?

There are multiple reasons for wanting to transform unstructured BPMN models into structured ones. Firstly, it has been empirically shown that structured process models are easier to comprehend and less error-prone than unstructured ones [1]. Thus, a transformation from unstructured to structured process model

J. Mendling, M. Weidlich, and M. Weske (Eds.): BPMN 2010, LNBIP 67, pp. 1–7, 2010.

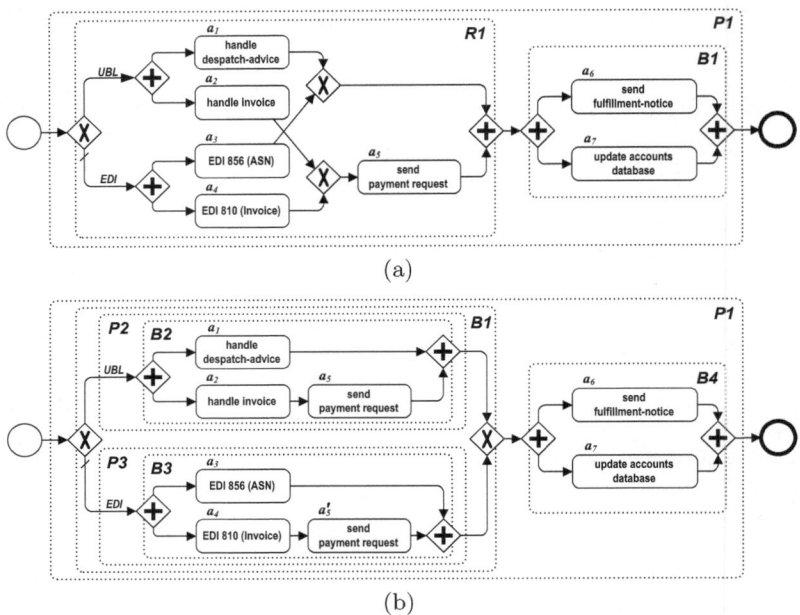

Fig. 1. Unstructured process model and its equivalent structured version

can be used as a refactoring technique to increase understandability. In particular, models generated by process mining techniques are often large, spaghetti-like and difficult to understand and would benefit from being re-structured. Also, mined process models come without layout information, thus requiring automated layout techniques to be applied. Automatic layout of structured process models is easier compared to layout of arbitrarily unstructured models.

Secondly, several existing process model analysis techniques only work for structured models. For example, an efficient method for calculating cycle time and capacity requirements for process models is outlined by Laguna & Marklund [2], but this method is only applicable to well-structured models. Other methods for computing the Quality of Service (QoS) of process models and service orchestrations assume that models are structured [3,4], and the same applies to methods for analyzing time constraints in process models [5]. By transforming unstructured process models to structured ones, we can extend the applicability of these analysis techniques to cover a larger class of models.

Finally, a transformation from unstructured to structured process models can be used to implement converters from graph-oriented process modeling languages like BPMN to structured process modeling languages such as BPEL [6].

3   A Short History of Structured Process Models

In many ways, *flowcharts* can be seen as predecessors of business process modeling notations such as BPMN. Thus, before discussing how to structure BPMN models, it is useful to summarize some key results of a large body of research that has tackled the problem of structuring flowcharts. This research, dating mostly from the 70s and 80s, was initially motivated by the debate between proponents of structured programming (based on "while" loops) and those who wanted to stick to programs with GOTO statements. Proponents of structured programming showed that any unstructured flowchart (representing a program with GOTO statements) can be transformed into a structured one. In fact, the title of the present paper is inspired by that of a seminal paper by Oulsnam [7], which presented a classification of unstructured flowchart components and showed how each type of component can be transformed into an equivalent structured one. Oulsnam and others noted that, when structuring loops with multiple exit points as well as overlapping loops, one needs to introduce boolean variables in order to encode parts of the control flow. In any case, we can retain from this work that every unstructured BPMN model composed of tasks, events, exclusive gateways and flows can be transformed into a structured BPMN model.

Another heritage from the research on program structuring is the *Program Structure Tree* (PST). The PST of a program is a tree in which the nodes represent SESE regions in the program's flowchart. The root of the PST represents the entire program. As we go down the PST, we find smaller SESE regions, until we reach individual steps. The SESE region associated to a node contains the SESE regions associated to each of its child nodes, and the SESE regions of these child nodes are disjoint. The concept of PST can be applied to BPMN models because it is unimportant whether the nodes represent tasks, events, exclusive or parallel gateways, or other BPMN nodes. Fig.1(a) shows the SESE regions composing the PST of the BPMN model. The largest region (P1) contains the entire BPMN model. Nested inside it we find two other regions (R1 and B1).

Recent research on structuring business process models has motivated further developments around the concept of PST. Motivated by the problem of transforming unstructured BPMN models into structured ones, Vanhatalo et al. [8] have proposed an improved version of the PST called the RPST (Refined PST). The RPST addresses some technical issues in the PST that we do not discuss because they are irrelevant to this paper. In fact, the RPST of Fig.1(a) and 1(b) are exactly the same as the corresponding PSTs. The difference between RPST and PST is only visible in more specific examples, particularly when some gateways are used both as split and joins.

The RPST also introduces a classification of SESE regions (also called *components*) into four classes: A *trivial* (*T*) component consists of a single flow arc. A *polygon* (*P*) represents a sequence of components. A *bond* (*B*) stands for a set of components that share two common nodes (basically: a split gateway and a join gateway). Finally, any other component that does not fall in these categories is a *rigid* (*R*) component. In Figures 1(a) and 1(b), the labels of the components reflect their types (e.g. *R1* is a rigid, *P1* is a polygon). For the purposes of this paper, trivial components are unimportant, and therefore we ignore them.

A process model is structured if its RPST does not contain any *rigid* component. For example, Fig.1(b) only contains *B* and *P* components. The problem of structuring BPMN model boils to transforming R components into combinations of P and B components.

The methods for structuring rigids differ depending on the types of gateways in the rigid and whether the rigid contains cycles or not. Accordingly, we classify rigids as follows (cf. Fig.2). A homogeneous rigid contains either only exclusive (*xor*) or only parallel (*and*) gateways. We call these rigids *(homogeneous) and rigids* and *(homogeneous) xor rigids*, respectively. A heterogeneous rigid contains a mixture of *and/xor* gateways. Heterogeneous and homogeneous *xor* rigids are further classified into cyclic or acyclic. We leave *cyclic homogeneous and rigids* out of the discussion, because it can be shown that

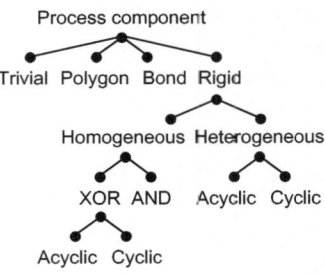

Fig. 2. Taxonomy

BPMN models containing such rigids are not *sound* according to the usual definition of soundness [9]. Soundness is a widely-accepted correctness criterion for process models.

One of the earliest studies on the problem of structuring BPMN-like models is that of Kiepuszewski et al. [10]. The authors showed that not all acyclic *and* rigids can be structured by putting forward a counterexample, which essentially boils down to the one in Fig.3. The authors showed that there is no structured (BPMN) model that is equivalent to this one under an equivalence notion known as Fully-Concurrent Bisimulation (FCB). This equivalence notion is arguably the

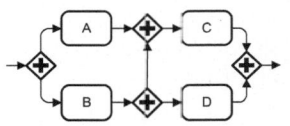

Fig. 3. Inherently unstructured BPMN model

one we seek in the context of transforming unstructured process models into a structured one. We could transform the model in Fig.3 into a trace-equivalent or weakly-bisimilar structured model that only contains exclusive gateways or event-driven decision gateways by enumerating all possible sequential executions of the tasks in the model, but this leads to spaghetti models. If there is parallelism in the original model, we also want the restructured model to have parallelism to the same extent. This is precisely what FCB-equivalence captures.

Liu & Kumar [11] continued this work by outlining a taxonomy of unstructured process model components. Their taxonomy puts in evidence several types of cyclic and acyclic rigids, distinguishing those that are sound and those that are not. The taxonomy includes a class of heterogenous acyclic rigids called *overlapping structures*, of which Fig.1(a) is an exemplar. Another example of

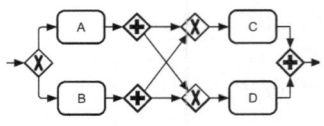

Fig. 4. Overlapped structure

an overlapped structure is shown in Fig.4. The authors note that such "overlapping structures" are sound and that they have an equivalent structured model, but without defining an automated method for structuring these and other unstructured rigids. Also, the taxonomy is not complete: some unstructured components do not fall into any of the categories. Later, Hauser et al. [12] outlined another classification of process components using *region trees* – a structure similar to the RPST. The authors showed a method for detecting and refactoring the "overlapped structures" identified by Liu & Kumar. Hauser et al. also observe that all homogeneous rigids are sound. Unsoundness comes from heterogeneous rigids.

We retain from the above that:

- Thanks to the RPST, we can structure a process model if we can structure every rigid component in the process model.
- Any homogeneous *xor* rigid can be structured (cf. GOTO-to-While problem).
- Some homogeneous *and* rigids cannot be transformed into equivalent structured components under FCB-equivalence.
- Heterogeneous rigids are unsound in some cases. When they are sound, it may or may not be possible to transform them into structured components.

4   Towards a Complete Structuring Method

In recent work [13], we presented a method for structuring acyclic BPMN process models. This method is implemented in a tool called BPStruct.

To intuitively explain how BPStruct works, we observe that when transforming an unstructured model into a structured one, we need to duplicate some tasks. For example, in Fig.1(b), task send payment request appears twice, whereas it only appears once in Fig.1(a). If we dig deeper, we observe that this duplication occurs when the unstructured model contains an *xor*-join that is not paired with a unique *xor*-split. This is the case for example in Figure 1(a), which features two *xor*-joins that are not paired with an *xor*-split. In this case, we need to duplicate the tasks that come after such an *xor*-join, and at the same time, push the *xor*-join downstream in the process model, until we get to a point where the *xor*-join is paired with an *xor*-split. In the general case, this "duplicate-and-push" procedure is complicated. But fortunately, this problem has been tackled in the context of Petri nets. Petri net researchers have developed techniques to "unfold" a net so that *xor*-joins are pushed as far as possible downstream. If we push the unfolding to the extreme, we obtain something called an occurrence net, which is basically a net without *xor*-joins.[1] While unfoldings solve the problem of duplicating tasks and getting rid of unstructuredness caused by improperly paired *xor*-splits, they can become quite large if we don't stop unfolding at the right point. Esparza et al. [14] have devised a technique that computes an unfolding that is rather small compared to other possible unfoldings. This is called the *complete prefix unfolding*, and it is the intermediate structure that BPStruct uses for structuring both acyclic and cyclic rigids.

In addition to duplicating tasks when required, an unfolding puts into evidence the fundamental ordering relations between pairs of activities. Specifically, the

---

[1] For those familiar with Petri nets, an *xor*-join translates into a place with two input arcs. A net in which every place has only one input arc is called an occurrence net.

unfolding allows us to easily determine which pairs of tasks are in a causal relation (meaning that the execution of one task causes the execution of another task), which pairs tasks are in a conflict relation (meaning that if one task is executed the other one will not) and which pairs of tasks are in a concurrency relation, meaning that they are both performed, but in any order. In other words, from the unfolding we can directly compute a graph of ordering relations between activities. Each edge in this graph is labelled by one of three types of relations: causality, conflict and concurrency. For example, the ordering relations extracted from the unfolding of the BPMN model in Fig.4 are shown in Fig.5(a). The filled one-way arrow denotes causality, the filled two-way arrow denotes conflict, and the dotted two-way arrow denotes concurrency.

The ordering relations capture all the control-flow information in the original model in a compact way. At this stage, all the necessary duplication of tasks has been done, and the model has been cleaned from spurious gateways that sometimes appear in unstructured models. In principle, we should be able to reconstruct a BPMN model by converting these ordering relations into flows and gateways. But how do we ensure that the resulting model is structured? Here is where another theory comes in very handy: that of *modular decomposition* of graphs. The modular decomposition theory enables us to find blocks in an arbitrary graph. BPStruct applies a modular decomposition algorithm on the graph of ordering relations in order to separate it into blocks. For example, the modular decomposition of Fig.5(a) is shown in Fig.5(b). Here we can clearly see that tasks a and b belong to a conditional block, because they are in conflict. Meanwhile, tasks c and d belong to a parallel block because they are in parallel. These two blocks are in a causality relation. This modular decomposition can then be used to synthesize the structured BPMN model shown in Fig.5(c).

We have shown in [13] that an acyclic rigid can be structured if its modular decomposition is composed of linear and complete modules as in the Fig.5(b). These modules basically correspond to P and B components in the RPST. If the ordering relations graph contains a third type of module known as a *primitive*, then the original process model is inherently unstructured. For example, Fig.5(d) shows the ordering relations graph of Fig.3. The modular decomposition of this graph contains only one primitive module. Hence, it cannot be structured.

The method outlined above only works for acyclic rigids because the graph of ordering relation is not a convenient abstraction for models with cycles. In the presence of cycles, we may have causal relations from task A to B and vice-versa,

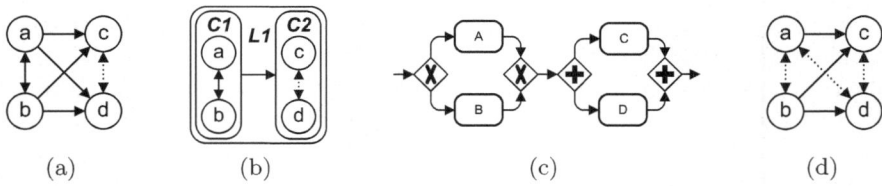

Fig. 5. (a) Ordering relations of Fig.4, (b) modular decomposition, (c) resulting structured component, (d) ordering relations of Fig.3

leading to a chicken-or-egg issue. Still, Petri net unfoldings can also be used to structure cyclic rigids. The idea is to extract acyclic parts from the unfolding, abstract them as black-boxes (that can be structured using the above method), and then construct a cyclic rigid that only contains *xor* gateways. This latter rigid can be structured using GOTO-to-While transformations. A preliminary solution for structuring cyclic rigids is implemented in BPStruct. However, as of the time of writing this paper, the underlying theory is still being worked out.

Also, BPStruct currently does not deal with inclusive and complex gateways, error events, exception flows, attached events and non-interrupting events. As the tool matures, we hope to lift as many of these restrictions as possible.

Acknowledgments. This work is supported by ERDF via the Estonian Centre of Excellence in Computer Science and the EU FP7 Project 257593 – ACSI.

References

1. Laue, R., Mendling, J.: The Impact of Structuredness on Error Probability of Process Models. In: UNISCON. LNBIP, vol. 5, pp. 585–590 (2008)
2. Laguna, M., Marklund, J.: Business Process Modeling, Simulation, and Design. Prentice-Hall, Englewood Cliffs (2005)
3. Cardoso, J., Sheth, A., Miller, J., Arnold, J., Kochut, K.: Quality of service for workflows and web service processes. Web Semantics: Science, Services and Agents on the World Wide Web 1(3), 281–308 (2004)
4. Zeng, L., Benatallah, B., Ngu, A.H.H., Dumas, M., Kalagnanam, J., Chang, H.: QoS-aware middleware for web services composition. IEEE Transactions on Software Engineering 30(5), 311–327 (2004)
5. Combi, C., Posenato, R.: Controllability in Temporal Conceptual Workflow Schemata. In: Dayal, U., Eder, J., Koehler, J., Reijers, H.A. (eds.) BPM 2009. LNCS, vol. 5701, pp. 64–79. Springer, Heidelberg (2009)
6. Ouyang, C., Dumas, M., van der Aalst, W.M.P., ter Hofstede, A.H.M., Mendling, J.: From business process models to process-oriented software systems. ACM Trans. Softw. Eng. Methodol. 19(1) (2009)
7. Oulsnam, G.: Unravelling unstructured programs. Comput. J. 25(3), 379–387 (1982)
8. Vanhatalo, J., Völzer, H., Koehler, J.: The Refined Process Structure Tree. Data & Knowledge Engineering 68(9), 793–818 (2009)
9. Kiepuszewski, B., ter Hofstede, A.H.M., van der Aalst, W.M.P.: Fundamentals of Control Flow in Workflows. Acta Inf. 39(3), 143–209 (2003)
10. Kiepuszewski, B., ter Hofstede, A.H.M., Bussler, C.: On Structured Workflow Modelling. In: Wangler, B., Bergman, L.D. (eds.) CAiSE 2000. LNCS, vol. 1789, pp. 431–445. Springer, Heidelberg (2000)
11. Liu, R., Kumar, A.: An Analysis and Taxonomy of Unstructured Workflows. In: van der Aalst, W.M.P., Benatallah, B., Casati, F., Curbera, F. (eds.) BPM 2005. LNCS, vol. 3649, pp. 268–284. Springer, Heidelberg (2005)
12. Hauser, R., Friess, M., Küster, J.M., Vanhatalo, J.: An Incremental Approach to the Analysis and Transformation of Workflows Using Region Trees. IEEE Transactions on Systems, Man and Cybernetics, Part C 38(3), 347–359 (2008)
13. Polyvyanyy, A., García-Bañuelos, L., Dumas, M.: Structuring acyclic process models. In: Proc. 8th International Conference on Business Process Management, Hoboken, NJ, USA (September 2010)
14. Esparza, J., Römer, S., Vogler, W.: An Improvement of McMillan's Unfolding Algorithm. FMSD 20(3), 285–310 (2002)

BPEL vs. BPMN 2.0: Should You Care?

Frank Leymann

Universität Stuttgart, Institute of Architecture of Application Systems (IAAS),
Universitätsstr. 38, 70569 Stuttgart, Germany
Leymann@iaas.uni-stuttgart.de

Abstract. BPMN 2.0 is an executable process modeling language. Thus, its relation to BPEL becomes an issue. In this paper, we propose a positioning of both languages, introduce the notion of a "native metamodel", and emphasize the role of the native metamodel of a process engine for the actual discussion.

Keywords: Workflow management systems, process engines, process modeling, metamodels, standards, BPEL, BPMN.

1 Introduction

In his blog post about "Which is simpler: BPMN or BPEL?" [8], Michael Rowley kicked-off another round of discussion about the use of BPEL vs the use of BPMN. Bruce Silver responded in his blog post "BPMN vs BPEL: Are we still debating this?" [10], which in turn triggered Michael Rowley to his post "BPMN 2.0 with BPEL - the debate is just starting" [9]. The author of this paper contributed his position on this subject in his post "BPMN 2.0 vs BPEL: Should you care?" [5]. Finally, Matthias Kloppmann commented on all these posts in "BPMN and BPEL – the relationship of two BPM standards" [4]. This paper is an extension of the post [5] of the author.

2 Positioning the Two Standards

The debate about the positioning of both standards is ongoing and it is an important debate to have, since the advent of BPMN 2.0 bares a lot of chances to confuse users of BPM technology about appropriateness of standards. The author is both, one of the original authors of BPEL [1], and also one of the authors of the submitted BPMN 2.0 specification [3]. Thus, somehow the author lives in both worlds.

We will argue that this is in fact one world or – more precise – two sides of one coin. This paper should help avoiding confusions about the role of both, BPMN 2.0 and BPEL. More background on BPM standards and their relations can be found in [7].

2.1 The Key Aspect of BPEL

The most important aspect of BPEL that cannot be emphasized enough is that BPEL exists, and that it is supported by most BPM middleware vendors in their products.

J. Mendling, M. Weidlich, and M. Weske (Eds.): BPMN 2010, LNBIP 67, pp. 8–13, 2010.

This results in both, the portability of skills as well as the interoperability of process models between tools. This is what Michael Rowley points out [8], and that we want to repeat here in order to emphasize this key point. Former attempts to standards in this area [7] did not achieve this at such a broad scale. Thus, BPEL is an important cornerstone of BPM technology. Especially, BPEL became the BPM runtime standard. This is in sharp contrast to the position Bruce Silver takes in [10].

BPEL focuses on providing a language for specifying business processes and provides a precise operational semantics for executing processes specified in this language. In doing so, BPEL combines two different approaches to specify business processes, namely a programmatic (or block-oriented) approach (stemming from XLANG) as well as a graph-oriented approach (stemming from WSFL) [7]. Especially, BPEL in fact is graph-oriented despite the many other claims: for example Bruce Silver in his post [10] is on the same wrong path and Michael Rowley gives an example of BPEL's graph-orientation [9]. The combination of these two approaches (block-oriented as well as graph-oriented) in fact is of utmost importance: both approaches had been supported by products at the time the work on BPEL began. Without their combination in BPEL the chasm between the two approaches and corresponding product worlds would have deepened – with obvious impact on the industry and users of BPM technology.

2.2 BPMN 1.x Filled a Hole

By focusing on language aspects and operational semantics aspects of business processes, BPEL left the visual representation dimension of such process models completely out of scope – by will, focusing on "time to market". A mistake when looking back in time from today.

This visual dimension became the domain of BPMN 1.x which now fills this gap. By providing a visual modeling language for business processes, BPMN 1.x enables non-IT experts to communicate and mutually understand their models: a big progress in this area resulting in the wide-spread use of BPMN 1.x.

2.3 The Chasm between BPMN 1.x and BPEL

Business processes specified in BPEL are nearly always modeled with their future execution in mind. More and more, processes modeled in BPMN are also not only modeled for documentation purposes but for purpose of their execution. Because of the wide-spread support of BPEL by middleware vendors the use of BPEL engines for BPMN execution is an obvious choice. To enable the execution of BPMN process models in BPEL engines, the transformation of BPMN process models to BPEL process models is needed. But the metamodel underlying BPMN and the metamodel underlying BPEL are not identical; in fact they are quite different. Thus, the transformation is not straightforward and sometimes requires complex mappings resulting in a BPEL process model that look quite different from the original BPMN process model (quite a number of publications exist that dive deep into this).

Consequently, we are faced with the following problem: How can we bridge the gap between BPEL (an execution language with rigor operational semantics) and BPMN 1.x (a visual language with informal semantics).

2.4 Options to Bridge the Chasm

One possible way to solve this problem would be to standardize a visual representation for BPEL. This is possible as proven by the many tools providing a graphical editor for BPEL process models. But the problem is that BPMN 1.x contains quite a number of constructs that are used by BPMN modelers in practice (e.g. arbitrary backward loops, gateways) and that these constructs have no direct correspondence in BPEL. Thus, BPEL extensions would be needed to support these constructs to allow for immediate seamless graphical representations as required by BPMN users. But extending BPEL is time consuming (e.g. the BPEL4People [2] extensions took five years from their original inception as a white paper in 2005 until the de jure standard in 2010) while BPMN users want to see a solution "soon". And, most importantly, following the way of creating a new visual language would introduce a new split in BPM technology – which must be avoided.

The other way to solve the problem – the way that has been taken – is to move BPMN 1.x forward in order to bridge the gap between BPMN and BPEL, i.e. to make transforming BPMN process models to BPEL much easier. The fundamental enhancements of BPMN 2.0 that enable this are (i) the addition of (visual) elements for BPEL handlers, (ii) the specification of a pattern-based approach of mapping a subset of BPMN 2.0 to BPEL, (iii) an XML-based exchange format, and most importantly (iv) the definition of an operational semantics for BPMN 2.0 that is close to BPEL (for those parts that correspond to BPEL). Effectively, BPMN 2.0 includes a subset of language elements that represent a visual language for BPEL and that can be naturally executed by a BPEL engine. This is a huge progress from a practical point of view.

2.5 The Price to Pay

Note, that it is a subset of BPMN 2.0 only that is naturally supported by BPEL. And, Michael Rowley is right that BPMN 2.0 has increased in terms of complexity [8]. Partially, this increase in complexity is owing to support the execution of BPMN 2.0 process models, which required precision in terms of type systems, service models etc. And BPMN 2.0 had to add new features required by domain users (most notably choreographies, collaborations, and conversations) – which are the other cause for the increase of complexity in BPMN 2.0. The latter kind of features is not mapped to BPEL by the BPMN 2.0 specification. This is because some of these BPMN features are out scope of BPEL (like BPMN conversations), and for some other features (like choreographies) it is still under discussion whether or how they should be supported by BPEL at all (like BPMN choreographies).

2.6 The Resulting Situation in Process Modeling

From a BPEL perspective the current situation is as follows: A subset of BPMN 2.0 is "isomorphic" to BPEL. As a consequence, BPMN 2.0 encompasses a visual modeling language for BPEL – this subset can be naturally transformed to BPEL and executed in a BPEL engine. From an architectural point of view, a tool supporting this BPMN 2.0 subset is a layer on top of BPEL engines – just like a data modeling tool supporting an entity-relationship model or an object model can be seen as a layer on top of relational database systems.

From a BPMN 2.0 perspective, the situation is as follows: BPMN 2.0 is a process modeling language with an operational semantics imprinted by BPEL (i.e. BPMN 2.0 builds on the success of BPEL). Thus, it is now possible to build an engine that directly supports BPMN 2.0 – without the intermediate step of generating BPEL. In other words, no BPEL at all is required to execute process models specified in BPMN 2.0. Just like database systems have been build in the past that support entity-relationship models or object models directly without the need for using relational database systems (this analogy is refined below in section 3.5).

3   Native Metamodel of a Workflow System

Effectively, we now have two standardized metamodels supporting the specification and execution of processes. And we will very likely see workflow systems (aka process engines) supporting one or the other metamodel, or even both [4]. The question to ask is: Why should (domain) users care about the metamodel of an engine anyhow?

3.1   The Notion of a Native Metamodel

Today, nearly nobody really cares about the metamodel of a BPEL engine at all! It might be surprising, but many BPEL engines in fact do not support BPEL "natively". But what does "natively" mean?

What characterizes a process engine at first sight as a *BPEL engine* is that it supports importing a process model specified in BPEL and executing it afterwards according to BPEL's operational semantics. In course of that, it typically digests the BPEL file and generates internal artifacts from it (e.g. it disperses the various BPEL elements across different tables; it creates Java representations from it; etc.) [6]. These internal artifacts, how they relate, and how they and their relations are interpreted by the implementing engine are what we suggest to call the engines *native metamodel.*

3.2   The Native Metamodel of a BPEL Engine

The native metamodel of a BPEL engine might be quite different from the BPEL metamodel. What determines the native metamodel of an engine is what the engineers building the engine chose to ensure efficiency, scalability etc of the execution.

Admittedly, the modeling language(s) to be supported at the time of the creation of the first release of a process engine has deep impact on the native metamodel of the engine. But good engine engineers take care about generality and extensibility of the native metamodel striving towards "straightforward" support of extensions of the original modeling language(s) and even new modeling languages. Typically, users of a BPEL engine are not aware of the engine's native metamodel and how different it is from BPEL as a metamodel itself.

3.3 Native Support of BPMN 2.0 – You Should Not Care

Similarly, users should not care about the native metamodel of an engine executing their BPMN 2.0 process model: whether the BPMN 2.0 process model is executed by a BPEL engine or by a new *BPMN* 2.0 *engine* is not important. In analogy to BPEL engines, the native metamodel of a BPMN 2.0 engine is likely to be different from BPMN 2.0 as a metamodel anyhow. I.e. the BPMN process models that are imported into such a BPMN engine will be transformed into a representation according to the native (and hidden) metamodel of that engine.

Note, that the native metamodel of a BPMN 2.0 engine might be one that supports also BPEL today. I.e. a vendor of a BPEL engine that provides import capabilities for BPMN 2.0 process models might become a vendor of a BPMN 2.0 engine based on the very same execution engine. Note, that it is absolutely valid that an intermediate BPEL process model from the BPMN 2.0 process model is generated before the actual import takes place – users might not even be aware of this intermediate BPEL format at all.

3.4 Key Aspects of a Process Engine

Far more important than the native metamodel of a process engine (especially a BPMN 2.0 engine) is its robustness, efficiency, scalability etc. Providers of today's BPEL engines typically have invested a lot into these non-functional properties of their engines. Thus, such vendors have the capabilities to also offer BPMN 2.0 engines with the very same non-functional properties.

3.5 Outlook

Finally, the analogy from the database domain sketched before is helping comprehension: When modeling data based on the object paradigm became important to domain users, database management systems have been built that supported the corresponding metamodel directly ("object oriented database systems"). But over the time existing relational database systems as well as the relational model itself have been extended to support key constructs of the object paradigm ("object-relational model").

Maybe the same will happen with existing BPEL engines and BPEL itself. I.e. BPEL might get extended over the time to support key features of BPMN 2.0 that are missing in BPEL and BPEL engines today. This is an interesting option, especially under the light of what Bruce Silver says, namely that it is likely that no vendor will support all of BPMN 2.0 in its product [10].

According to this opinion, subsetting of BPMN 2.0 will always happen and domain users will have to live with restricted support of BPMN 2.0 in corresponding modeling tools and runtime engines. Maybe this will give BPEL and those BPEL engines supporting "just" the BPEL-isomorphic BPMN 2.0 subset time to catch up to finally support the relevant missing elements of the BPMN 2.0 metamodel.

Acknowledgments. I am very grateful to Matthias Kloppmann for the many discussions we had on this topic; nevertheless, all opinions expressed in this article are solely the author's opinions.

References

1. Web Services Business Process Execution Language Version 2.0, OASIS (2007),
   `http://docs.oasis-open.org/wsbpel/2.0/wsbpel-v2.0.pdf`
   (June 20, 2010)
2. WS-BPEL Extension for People (BPEL4People) Specification Version 1.1,
   `http://docs.oasis-open.org/bpel4people/bpel4people-1.1.html`
   (June 20, 2010)
3. Business Process Model and Notation, OMG 2009 (2009),
   `http://www.omg.org/spec/BPMN/2.0/Beta1/PDF/` (June 20, 2010)
4. Kloppmann, M.: BPMN and BPEL – the relationship of two BPM standards,
   `https://apps.lotuslive.com/bpmblueworks/blog/?p=889` (June 20, 2010)
5. Leymann, F.: BPEL vs BPMN 2.0: Should you care?
   `http://leymann.blogspot.com/` (June 20, 2010)
6. Leymann, F., Roller, D.: Production Workflow. Prentice-Hall, Englewood Cliffs (2000)
7. Leymann, F., Karastoyanova, D., Papazoglou, M.: Influential BPM Standards: History and
   Essence. In: vom Brocke, J., Rosemann, M. (eds.) Handbook on Business Process Management. Springer, Heidelberg (2010)
8. Rowley, M.: Which is simpler: BPMN or BPEL,
   `http://www.vosibilities.com/bpel/`
   `bpmn-or-bpel-which-is-simpler/2009/11/19/` (June 20, 2010)
9. Rowley, M.: BPMN 2.0 with BPEL – the debate is just starting,
   `http://www.vosibilities.com/bpel/bpmn-with`
   `-bpel-the-debate-is-just-starting/2009/11/23/` (June 20, 2010)
10. Silver, B.: BPMN vs BPEL: Are we still debating this?
    `http://www.brsilver.com/2009/11/19/`
    `bpmn-vs-bpel-are-we-still-debating-this/` (June 20, 2010)

An Overview of BPMN 2.0 and Its Potential Use

Hagen Völzer

IBM Research – Zurich

BPMN (Business Process Model and Notation) is an OMG standard for business process modeling that is widely adopted today. The OMG lists 62 tool vendors who offer products that support BPMN [3]. This success is based on the fact that BPMN provides a standardized graphical notation which is easy to use for business analysts, allowing them to document and communicate their business processes within their company and with their external business partners.

BPMN was introduced in 2002, standardized by the OMG in version 1.0 in 2006 [1] and version 2.0 [2] is currently being finalized by the OMG. Version 2.0 takes BPMN to a new level. It adds the following to the previous version:

- a standardized metamodel and serialization format for BPMN, which allows users to exchange business process models between tools of different vendors,
- a standardized execution semantics for BPMN, which allows tool vendors to implement interoperable execution engines for business processes,
- a diagram interchange format, allowing users to exchange graphical information of a business process diagram,
- an extended notation for cross-organizational interactions (also known as *process choreographies*), which enables new use cases for automated tool support for processes that involve several business partners,
- a detailed mapping from BPMN to WS-BPEL, which demonstrates the alignment of BPMN with existing tools and standards, and
- some additional modeling elements for processes such as non-interrupting events and event subprocesses.

Among the many benefits of these additions, two are noteworthy in particular. Firstly, the standardized interchange supports collaboration between different organizations within and across different enterprises. A BPMN diagram created in some part of a company can be corrected, refined, complemented, analyzed or executed somewhere else using different tools of possibly different tool vendors.

Secondly, BPMN 2.0 is the first notation and interchange format that combines business user-friendly modeling with the detailed technical specification of an executable model within the same process model. This fosters collaboration between business analysts and technical developers of the IT system that supports the business. Together with the increasingly common (web-based) business process collaboration tools, this enables more agile approaches to the development and adaptation of information systems.

BPMN 2.0 defines a rich notation and modeling language that can be used in different ways and different scenarios. For example, one company could use BPMN to model their internal processes in order to analyze and improve them. Another company could use BPMN to model their interaction with business

J. Mendling, M. Weidlich, and M. Weske (Eds.): BPMN 2010, LNBIP 67, pp. 14–15, 2010.

partners within a particular supply chain. A third company might use BPMN to define their fully or partially automated production workflow that is executed by a powerful execution engine.

In the talk, we give an overview of such scenarios and the corresponding modeling elements that are used in these scenarios. Not all of these scenarios will be supported by every BPMN tool. This is reflected by the fact that the BPMN 2.0 specification provides different conformance classes to suit specific scenarios. The specification [2] defines Process Modeling Conformance, Process Execution Conformance, BPEL Process Execution Conformance and Choreography Modeling Conformance. Process Modeling Conformance is subdivided along different classes of process diagrams, into Descriptive, Analytic and Common Executable. In the talk, we also relate these conformance classes to the various uses of BPMN.

References

1. OMG. Business process modeling notation (BPMN) version 1.0, OMG document number dtc/06-02-01 (2006)
2. OMG. Business process model and notation (BPMN) version 2.0, OMG document number dtc/2010-05-03 (2010)
3. OMG, Business Process Management Initiative, BPMN Implementors Listing, http://www.bpmn.org/BPMN_Supporters.htm (accessed on June 2, 2010)

BPMN 2.0 Execution Semantics Formalized as Graph Rewrite Rules

Remco Dijkman and Pieter Van Gorp

Eindhoven University of Technology, The Netherlands
{r.m.dijkman,p.m.e.v.gorp}@tue.nl

Abstract. This paper presents a formalization of a subset of the BPMN 2.0 execution semantics in terms of graph rewrite rules. The formalization is supported by graph rewrite tools and implemented in one of these tools, called GrGen. The benefit of formalizing the execution semantics by means of graph rewrite rules is that there is a strong relation between the execution semantics rules that are informally specified in the BPMN 2.0 standard and their formalization. This makes it easy to validate the formalization. Having a formalized and implemented execution semantics supports simulation, animation and execution of BPMN 2.0 models. In particular this paper explains how to use the formal execution semantics to verify workflow engines and service orchestration and choreography engines that use BPMN 2.0 for modeling the processes that they execute.

1 Introduction

The Business Process Modeling Notation (BPMN) version 2.0 [1] has a well defined execution semantics. The execution semantics can be used by tool vendors to develop tools for simulation, animation and execution of business process models.

Tool developers must strictly adhere to the execution semantics, because interchange of business process models between tools is only possible if both syntax *and* semantics are preserved. In particular business process analyst can model a business process in one tool and expect the modeled process to behave in a certain way because they are familiar with the execution semantics of BPMN and/or because the modeling tool supports animation of the modeled process. If the modeled business process is subsequently exported to another tool, such as a workflow engine, the business process analyst expects the workflow engine to behave in the exact same way. The interchangeability of business process model semantics has become even more important now that interchange of syntax is facilitated well by the XML Process Definition Language (XPDL) [2], which is being adopted by an increasing number of tool vendors [3], making it more and more likely that process analysis use multiple tools during the business process lifecycle.

To further facilitate interchange of business process model semantics test suites are required to test whether a tool conforms to a specified execution

J. Mendling, M. Weidlich, and M. Weske (Eds.): BPMN 2010, LNBIP 67, pp. 16–30, 2010.

semantics. The use of test suites to test whether a tool conforms to a specified syntax is already common. For example, the XPDL specification contains a Document Type Definition (DTD) that can be used to verify whether the XPDL output that is generated by a tool conforms to the XPDL standard.

Therefore, the goal of this paper is to present a formalized execution semantics for a subset of BPMN. The execution semantics is defined used graph rewrite rules [4], which facilitate a strong traceability between BPMN execution rules and the formal semantics, thus enabling a formalized semantics that can easily be validated. As a proof of concept, the execution semantics is implemented in a graph rewrite tool called GrGen [5]. The formalized execution semantics is intended to be used as a test suite, to test conformance of tools that implements the BPMN execution semantics, in particular workflow engines and service choreography and orchestration engines. Therefore, this paper also presents an architecture to enable conformance testing of implementations of the BPMN execution semantics.

The remainder of this paper is structured as follows. Section 2 explains the concept of graph rewrite rules, which is used to formalize the BPMN execution semantics in this paper. Section 3 presents the formalization of the execution semantics. Section 4 presents a tool architecture that can be used, in combination with the execution semantics, to do conformance testing of tools that implement the BPMN execution semantics. Section 5 presents related work on defining the BPMN semantics and section 6 concludes.

2   Graph Rewrite Rules

Until the early 1970s, the semantics of programming languages was defined using tree-based structures. This followed from the use of context-free grammars for defining their textual syntax. Since diagrammatic languages did not always contain a clear tree basis, new techniques were clearly needed (as explained further by [6]).

In the meanwhile, significant theoretical and practical progress has been made and a dedicated conference (ICGT) is approaching its fifth edition. Various theories (such as the so-called double-pushout approach [7]) are actively being used and extended to reason about increasingly powerful language constructs for defining rewrite rules. On the practical side, the community holds a yearly contest involving various case studies for comparing the power of the supportive tools. The previous edition of that contest has indicated that graph transformation tools can successfully be used for re-implementing existing BPMN related transformation programs in a more concise manner [8].

In this paper, we describe how we are using one of the contest's most successful graph transformation tools on a new BPMN related case study. Although we do not aim to explain each individual GrGen language construct, this section should enable BPM researchers without previous experience with graph transformation to comprehend some fragments of our formalization and to explore further details by executing the related rewrite system in debugging mode. To facilitate the latter, we provide an online virtual machine containing all software related to this paper [9].

A graph rewrite rule is the basic building block of any graph transformation system. It consists of two conceptual parts: its left-hand side specifies a pattern defines the precondition for rule application whereas its right-hand side defines the pattern that should be realized when *applying* the rule on a match in the host graph. Graph rewrite rules can be specified both in a textual and a graphical manner. Figure 1 shows an example of a graph transformation rule. The rule is shown both using graphs and using the BPMN notation. The left-hand side and the right-hand side of the rule are shown on the left-hand and right-hand of the large arrow. The rule matches a graph in which there exists a node 'sf' of the type 'SequenceFlow' from which there is an edge named 't' of type 'Tokens' to a node 'tok' of the type 'Token' and an edge of type 'To' to a node 'a' of type 'Activity'. The rule realizes the deletion of the edge named 't' and the creation of an edge of type 'Tokens' from 'a' to 'tok'. In other words, if there is an activity with an incoming sequence flow with a token on it, the rule can move the token to the activity. If the rule invokes any other rules, this is graphically represented in the block below the rule. Below, the same information is represented textually.

```
1 rule enterNodeOneTrans {
2      tok:Token <-ts:Tokens- sf:SequenceFlow -:To->
              a:Activity;
3      modify{
4          delete(ts); a -:Tokens-> tok;
5      }
6 }
```

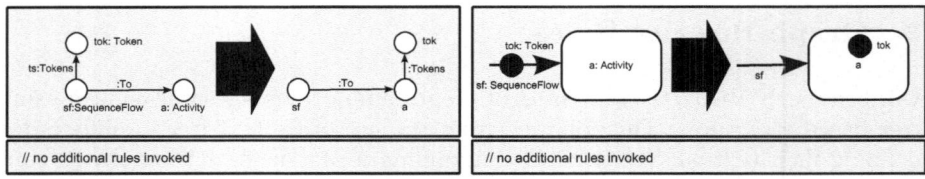

Fig. 1. Example of graphical representation of graph transformation rules

The GrGen pattern definition part (left-hand side) consists of node and edge declarations (or references to already declared elements). Nodes are declared by n:t, where n is an optional node identifier, and t its type. An edge e with type t and between nodes n1 and n2 is declared by n1 -e:t-> n2 (or n1 <-e- n2, or n1 -e- n2), whereas n1 --> n2 describes an anonymous edge. Nodes can also be left anonymous in patterns. The pattern "a:Activity --> . --> . --> . --> a" for example describes a circular structure connected to activity node a without mentioning irrelevant names for the intermediate node or edge variables.

All nodes and edges are implicit instances of Node and Edge respectively. Gr-Gen also offers a textual language for defining more specific node types (such as Activity in the above example) and edge types. Both nodes and edges can have attributes and edges do not need to be directed (as opposed to term rewriting or ECORE based approaches [10]). Nodes and edges are referenced outside their declaration by n and -e->, respectively.

The GrGen rewrite part (right-hand side) is specified by a block that is nested within the rule. In graph transformation theory, elements that are part of the right-hand side but no part of the left-hand side will be created upon rule application; elements that are part of the left-hand side but no part of the right-hand side will then be deleted. Although GrGen has a dedicated operator for this specification style, it also enables a variant where the deletion of elements needs to be encoded explicitly. In this paper, we only apply this variant. In the GrGen textual syntax, the rewrite part is then contained in a `modify` block. The GrGen user manual explains further details and also clarifies how patterns can be defined recursively, how matches can be executed in parallel, etc.

3 Formalization of the BPMN Execution Semantics

This section shows first how the rules for the BPMN execution semantics are controlled and then clarifies the meaning of some illustrative rules. As in the specification of the BPMN execution semantics, these rules are defined in terms of enabling and firing of elements, based on a token-game. The last subsection of presents the precise rules for moving tokens as well as keeping track of past markings. The complete execution semantics can be downloaded[1].

3.1 Execution Semantics Control

The "master script" of the GrGen implementation of the BPMN execution semantics is shown below.

```
1 include ../tests/Subprocess4.grbpmn       8       enterIntermediateEvent $||
2 validate strict                           9       enterExclusiveGateway $||
3 xgrs [mapTo_SequenceFlows] | [mapTo_Associations]   10      enterInclusiveGateway $||
       | addActivitiesToWFProcessWhereNeeded |       11      enterParallelGateway $||
       [addSequenceFlowToWFProcessOfFromActivity]    12      enterSubprocess $||
                                            13      leaveNodeOneTrans $||
4 xgrs [AddTokenForGlobalProcess] &&        14      leaveSubprocessNormal $||
       CreateInitialMarking                 15      leaveExclusiveGateway $||
                                            16      leaveParallelGateway $||
5 validate strict                           17      leaveInclusiveGateway
6 debug xgrs (                              18 ) [*]
7     enterNodeOneTrans $||
```

Lines 6 to 18 execute the BPMN execution semantics rules. As documented in the rule names, some rules realize the *entry* of tokens in activity nodes whereas other rules ensure that tokens can leave such nodes again. The `xgrs` construct enables the execution of rewrite rules. These rules can be controlled by operators such as [] (parallel rule application), || (sequential OR with lazy evaluation), etc. The $ operator flags on lines 6 to 18 indicates that the rules under consideration commute (they can be evaluated in any order). The * flag, which is used in `(rules)[*]`, instructs that the *rules* should be executed iteratively as long as matches are found.

Lines 1 through to 5 execute some initialization. Line 1 loads the input BPMN model and line 2 verifies whether the input graph conforms to the BPMN metamodel. Such syntactical checks have been quite useful during the concurrent

---

[1] Implementation available at: `http://is.tm.tue.nl/staff/rdijkman/bpmn.html`

development of the XPDL parser and the actual GrGen transformation. Line 3 transforms the input model in a form that is required to reason precisely about the state during BPMN model execution. More specifically, BPMN sequence flows and associations are transformed into pairs of edges with an intermediate node. This node can then be associated with the tokens that reside on that edge during process execution. Notice that this would not be necessary if GrGen would support so-called hyper-edges but we are in favor of GrGen's simple yet sufficiently powerful metamodeling foundation.

Line 4 creates the initial marking (i.e., the initial overall process state). It puts tokens in the appropriate activities and creates a process instance element for each top-level workflow process.

3.2   Execution Semantics Rules

In this subsection we explain the actual execution semantics rules that are listed in lines 6 to 18.

The first rule 'enterNodeOneTrans' represents the execution semantics rule for an activity with a single incoming sequence flow. It is graphically represented in figure 2 and listed below.

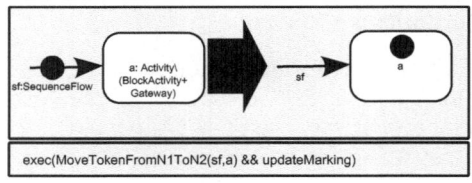

Fig. 2. Visual representation of rule *enterNodeOneTrans*

The left-hand side of the rule matches an element 'a' from the set of activities minus the set of block activities (i.e. embedded or referenced subprocesses) and gateways. The matched activity must have an incoming sequence flow 'sf' on which there is a token 'tok'. Note that this means that an activity can be a task, an intermediate event or an end event, although figure 2 shows a task. Also note that this rule does not require that an activity only has one incoming sequence flow; it also applies in case an activity has multiple incoming sequence flow and, therefore, implements the rule that an activity with multiple incoming sequence flows behaves as an exclusive join gateway. If such a match is found, the rule executes the right-hand side, which involves the execution of the rule that moves the token 'tok' from one place to another (in this case from the sequence flow 'sf' to the activity 'a'). Our implementation of the execution semantics stores the execution trace that was performed, this is done by executing the 'updateMarking' rule. The 'MoveTokenFromN1ToN2' and 'updateMarking' rules are used by all other rules. They are explained in subsection 3.3.

The second rule 'enterIntermediateEvent' represents the execution semantics rule for canceling a task or a subprocess, because an intermediate boundary event occurs. It is graphically represented in figure 3[2]. The BPMN 2.0 standard also allows for intermediate boundary events that do not cancel the activity to which they are attached. However, the current version of XPDL does not support the 'cancelActivity' flag. Therefore, we do not consider that flag and use intermediate activities that cancel the activity to which it is attached as standard behavior, because this was the standard behavior for previous versions of BPMN.

Fig. 3. Visual representation of rule *enterIntermediateEvent*

The third and fourth rule, 'enterExclusiveGateway' and 'enterParallelGateway', represent the execution semantics rules for multiple incoming sequence flows into an exclusive or parallel gateway, respectively. They are graphically represented in figure 4. The 'enterParallelGateway' rule uses a red box around a graph element. This red box represents that a graph is matched only if it *does not have* the construct in the red box. In this case the matched graph *does not have* a sequence flow to the gateway that *does not have* a token on it (i.e.: all sequence flows to the gateway must have a token on them).

Note that these execution semantics rules for the exclusive and parallel gateways also apply in case those gateways have multiple incoming sequence flows. Also note that, instead of executing the 'MoveTokenFromN1ToN2' rule, the parallel gateway executes the 'MoveTokenFromIncomingToN2' rule, because the tokens from all its incoming sequence flows must be moved instead of from a single incoming flow.

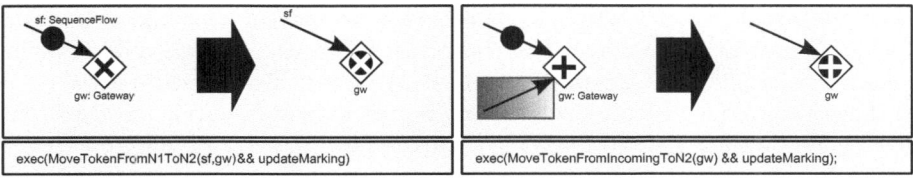

Fig. 4. Visual representation of *enterExclusiveGateway* and *enterParallelGateway* rules

---

[2] For the remaining rules, we only show the graphical representation. The listing can be downloaded.

The fifth rule 'enterInclusiveGateway' represents the execution semantics rule for multiple incoming sequence flows into an inclusive gateway. It is graphically represented in figure 5. According to the BPMN sepcification, an "inclusive gateway is activated if:

- At least one incoming sequence flow has at least one Token and
- for each empty incoming sequence flow, there is no Token in the graph anywhere upstream of this sequence flow, i.e., there is no directed path (formed by Sequence Flow) from a Token to this sequence flow unless
  - the path visits the inclusive gateway or
  - the path visits a node that has a directed path to a non-empty incoming sequence flow of the inclusive gateway." [1]

For each empty incoming sequence flow the rule 'HasNoTokenUpstream' checks whether that incoming sequence flow has no token upstream (for which the properties do not hold). We refer to the downloadable tool for the definition of this rule.

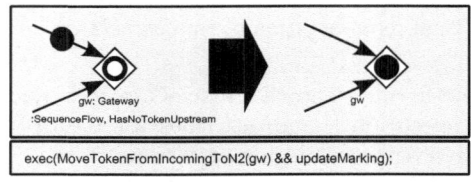

Fig. 5. Visual representation of rule *enterInclusiveGateway*

The sixth rule 'enterSubprocess' represents the execution semantics rule for instantiating a subprocess. This rule has two cases, one for the situation in which the subprocess contains a single start event and one for the situation in which the subprocess has no start event. These cases are illustrated in figure 6. In case the subprocess has a single start event, that event get a token when the subprocess is instantiated. In case the subprocess does not have a start event, activities and gateways that do not have incoming sequence flows and that are not the target of a compensation handler get a token when the subprocess is instantiated.

Tokens that are put on elements inside a subprocess instance should be 'colored' with that instantiation. This is necessary to prevent that tokens from different subprocess instances are somehow related, for example when determining whether a parallel join gateway is enabled. We leave this for future work.

The seventh rule 'leaveNodeOneTrans' represents the execution semantics rule for a node that has a single outgoing sequence flow. It is graphically represented in figure 7. Note that a node can in this case also be an intermediate event. We leave the case in which a task has multiple outgoing sequence flows for future work.

The eighth rule 'leaveSubprocessNormal' represents the execution semantics rule for completing a subprocess normally. It is graphically represented in figure 7. This rule has two cases, the case in which the subprocess has end events

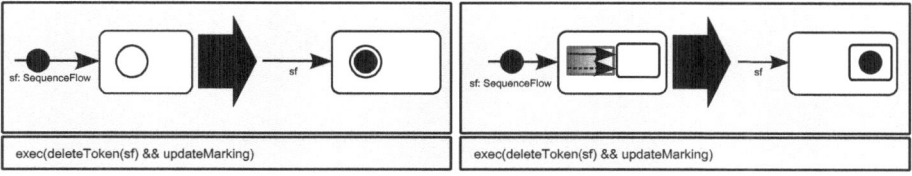

Fig. 6. Visual representation of rule *enterSubprocess*

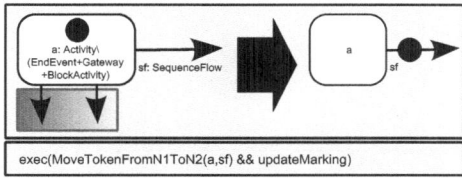

Fig. 7. Visual representation of rule *leaveNodeOneTrans*

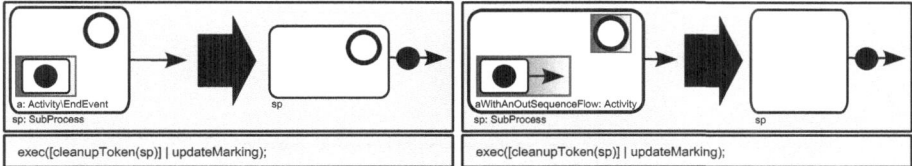

Fig. 8. Visual representation of rule *leaveSubprocessNormal*

and the case in which the subprocess does not have end events. In the first case the subprocess terminates when there is no token on an activity that is *not* an end event. In the second case the subprocess terminates when there is no token on an activity that has outgoing sequence flows. In both cases the sequence flow that leaves the subprocess receives a token and the tokens that still exist in the subprocess are cleaned up by the 'cleanupToken' rule.

The remaining rules apply to leaving gateways. Figure 9 shows the execution semantics rule for an exclusive split gateway. From an exclusive gateway with a token and an outgoing sequence flow, the token is removed from the exclusive gateway and placed on the sequence flow. Therewith, it creates exclusive split behavior, because the token can be put only on a single outgoing sequence flow. Figure 10 shows the execution semantics rule for a parallel split gateway. From a parallel gateway with a token, the token is removed. Subsequently, the rule 'enableOutflow' ensures that a token is put on each sequence flow that leaves the gateway. Figure 11 shows the execution semantics for an inclusive split gateway. This rules for this gateway are similar as those for the parallel split gateway.

Fig. 9. Visual representation of rule *leaveExclusiveGateway*

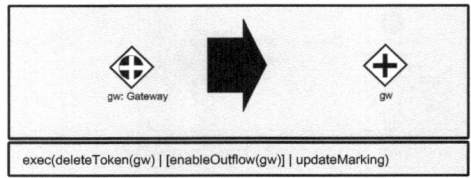

Fig. 10. Visual representation of rule *leaveParallelGateway*

Fig. 11. Visual representation of rule *leaveInclusiveGateway*

However, instead of putting tokens on all outgoing sequence flows, a token is put either on the default outgoing sequence flow or on some of the other outgoing sequence flows.

3.3 Moving Tokens

The execution semantics rules in the previous subsection use the 'MoveToken-FromN1ToN2' and 'updateMarking' rules to move tokens and keep track of previous markings.

The rule 'MoveTokenFromN1ToN2' is graphically represented in figure 12. This rule moves a token into a place (where a 'place' can both be a sequence flow and an activity) and deletes it from another place. Therewith, it effectively moves the token from the place in which it is deleted to the place on which it is put. The 'deleteToken' rule is shown in figure 13. It does not actually delete the token as its name implies, but rather changes the token into a 'hidden' token. Hidden tokens are tokens that cannot enable BPMN elements. We use them to keep track of past markings.

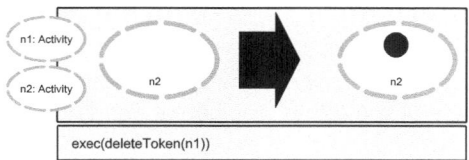

Fig. 12. Visual representation of rule *MoveTokenFromN1ToN2*

Fig. 13. Visual representation of rule *deleteToken*

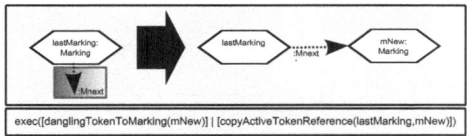

Fig. 14. Visual representation of rule *UpdateMarking*

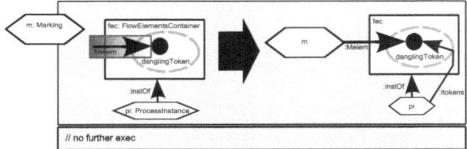

Fig. 15. Visual representation of rule *danglingTokenToMarking*

The other 'helper' rules are defined in a similar manner as 'MoveToken-FromN1ToN2' and 'deleteToken'. These rules are: 'MoveTokenFromIncoming-ToN2', 'cleanupToken', 'enableOutflow' and 'enableOutflowSome'.

The rule 'UpdateMarking' is graphically represented in figure 14. This rule finds the last marking in the execution trace. This is a marking that has no pointer 'Mnext' to the next marking. It then creates a pointer from that marking to a new marking. Subsequently it executes the rule 'copyActiveTokenReference', which creates references from the new marking to all tokens from the previous making that are active (i.e.: not hidden). Also, it executes the rule 'dangling-TokenToMarking', which creates references from the new marking to all tokens that are not yet referenced by a marking. These are the tokens that are created when firing the last BPMN execution rule. The 'danglingTokenToMarking' rule is graphically represented in figure 15.

4   Testing Conformance to Execution Semantics

The goal with which we develop the execution semantics is to use it as a means to check the conformance of tools that implement the BPMN execution semantics, such as workflow engines.

To this end we implemented the execution semantics described in the previous section in a tool called GrGen [5]. This tool can check whether a graph rewrite rule can be executed at a particular time and, if so, it can execute that rule. Since the rewrite rules represent execution semantics rules, this means that we can check whether an execution semantics rule can be executed at a particular time and, if so, we can execute that rule. Using the GrGen implementation of the execution semantics, we can verify conformance of a workflow engine to the BPMN execution semantics as shown in figure 16.

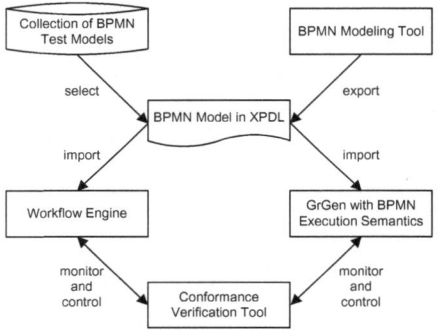

Fig. 16. Conformance Testing of Workflow Engines

The figure shows that we assume XPDL as the interchange format between BPMN modeling tools, workflow engines and the GrGen implementation of the execution semantics. Specifically, we use XPDL version 2.1, because it is the latest stable version of XPDL. The exact version that we use is important, because depending on the version of XPDL that we use, certain BPMN can or cannot be interchanged (in a standard manner). Consequently, the XPDL version that is used determines the BPMN constructs for which we can check the conformance. XPDL version 2.1 supports all sequence flow constructs within the scope of this paper, except for the parallel event-based gateway and the boundary intermediate event that *does not* cancel the activity to which it is attached (i.e. that has the attribute 'cancelActivity' set to 'false').

A BPMN model in XPDL can be imported both by a workflow engine (that supports XPDL) and our GrGen implementation of the execution semantics. A verification tool then tests the conformance of the workflow engine. It does this by monitoring and controlling the behavior of both the workflow engine and the execution semantics and verifying that the workflow engine changes state in the same manner as the execution semantics. To this end the verification tool does the following:

1. Determine the execution traces that can be performed by the execution semantics.
2. For each execution trace determine possible values for datafields and in which activity they must be entered.
3. Perform each execution trace in the workflow engine and check whether at each moment in the execution the set of activities that is allowed by the workflow engine is identical to the set of activities that is allowed by the execution semantics.

The implementation of the verification tool is left for future work.

We have also developed a collection of BPMN models to test conformance. Each of the BPMN models in this collection is developed to test a specific execution semantics rule that is defined in the BPMN standard. This collection of models can be used for unified testing, benchmarking and reporting on execution semantics conformance of different workflow engines. A conformance report can indicate specific rules that are or are not correctly implemented by the workflow engine.

5   Related Work

The BPMN 2.0 standard specifies the complete BPMN 2.0 execution semantics in natural language. The use of natural language is sufficiently precise to allow for an intuitive understanding of the execution semantics, but it cannot be directly implemented into a tool for purposes of simulation, verification or execution. Therefore more precise semantics for BPMN have been defined [11,12,13,14,15,16,17]. These semantics differ with respect to the means that is used to specify the semantics, the goal with which the semantics is specified and the conceptual focus of the semantics. Table 1 summarizes them.

Wong and Gibbons [11,12] define a semantics for a subset of the BPMN control-flow concepts in terms of the process algebra CSP [18]. This semantics allows them to check the consistency of business process models at different levels of abstraction (i.e. refinement checking). It also allows them to specify and check certain properties that must apply to the process. This includes domain specific properties, such as 'after an order is placed, a reponse must be sent to the client within 24 hours', and properties that apply to business process models in general, such as deadlock-freeness and proper completion [19]. We refer to the latter form of property checking as soundness checking. Dijkman et al. [13] define a semantics for a subset of the BPMN control-flow concepts in terms of classical Petri nets. The goal of their semantics is to define the BPMN execution semantics precisely and to enable soundness checking. Prandi et al. [14] define a semantics in terms of a process algebra called COWS [20]. Their semantics allows for soundness checking of BPMN models and also of quantitative simulation of BPMN models, provided that simulation information is provided with the BPMN model. The semantics is defined for a subset of both the control-flow and the data-flow aspect. Raedts et al. [15] define a semantics for a subset of the BPMN control-flow concepts in terms of classical Petri nets. The goal of their semantics is to enable soundness checking. Dumas et al. [16] define the execution semantics of a particular BPMN construct: the inclusive join gateway. Their goal is to discuss the execution semantics of this particularly complex construct in enough detail to allow animation of BPMN models that use this construct. Takemura [17] defines a semantics for the concepts that are related to BPMN transactions in terms of classical Petri nets. The goal of the semantics is to define the execution semantics of BPMN transactions precisely and to enable soundness checking.

The semantics in this paper differs from the other semantics with respect to the means that are used for the semantics and the prospective use of the semantics. This paper uses graph rewrite rules to define the semantics. One benefit of using graph rewrite rules is that a direct mapping is possible from the execution semantics rules in the BPMN specification to graph rewrite rules. This direct mapping makes the graph rewrite rules easily traceable to BPMN execution semantics rules and easily understandable. Another benefit of using graph rewrite rules is their relative expressive power. For example, classical Petri nets are inherently limited in the semantics that they can represent; it is notoriously hard to represent the OR-join in classical Petri nets and data-related concepts cannot be represented in a feasible manner in classical Petri nets. Such concepts can easily be represented in graph rewriting systems. We aim to exploit this expressive power in future work to define a complete formal semantics of BPMN. These properties enable the use of our semantics for conformance verification. In particular it enables us to compare the execution of a running workflow system to the execution semantics as it is executed in a graph rewriting tool.

Table 1. Semantics defined for BPMN

Semantics	Means	Uses	Focus
BPMN Standard [1]	Natural language	Semantics specification	Complete
Wong and Gibbons [11,12]	CSP	Refinement checking Property checking Soundness checking	Control-flow subset
Dijkman et al. [13]	Petri nets	Semantics specification Soundness checking	Control-flow subset
Prandi et al. [14]	COWS	Soundness checking Quantitative simulation	Control-flow subset Data-flow subset
Raedts et al. [15]	Petri nets	Soundness checking	Control-flow subset
Dumas et al. [16]	Pseudo code	Semantics specification	OR-Join
Takemura [17]	Petri nets	Semantics specification Soundness checking	Transactions
This paper	Graph rewriting	Semantics specification Conformance checking	Control-flow subset

6   Conclusions

This paper presents a formalization of a subset of the BPMN 2.0 execution semantics in terms of graph rewrite rules. The paper shows that there is a direct correspondence between the rules that informally define the BPMN 2.0 execution semantics in the standard and the graph transformation rules that formally define the BPMN 2.0 execution semantics in this paper. The benefit of having this direct correspondence is that the formalization can easily be validated, because the formal definition of each rule can be checked with its informal specification in the BPMN 2.0 standard and the completeness of the rules can be determined by checking whether the complete specification of the execution semantics is covered.

The formalization of the execution semantics in this paper can be used, among other things, for simulation, animation and execution of BPMN models. For this purpose it was implemented into a tool called GrGen. In particular we aim to use the formalization as a tool to verify the conformance of workflow engines that execute BPMN models to the BPMN execution semantics. To this end we will implement a monitoring and conformance verification tool in future work.

Another purpose for which formal semantics of process models are often used is for verification of correctness of those process models. The semantics that we defined in this paper is less suitable for this purpose, because it makes no claims about the statespace that can be constructed with it, such as whether the statespace is finite or whether the algorithm to verify the reachability of certain states can complete. However, formal semantics of BPMN in terms of Petri-nets or Process Algebra exist and can be used for that purpose.

It is the aim of this work to formalize the execution semantics of BPMN in its entirety, also covering the data and resource aspects of BPMN. To that end we will implement formal execution semantics rules for the data and resource aspect in future work. We will also complete the formal execution semantics for the control-flow aspect.

References

1. Object Management Group: Business process model and notation beta 1 for version 2.0. Technical Report dtc/2009-08-14, Object Management Group, Needham, MA, USA (2009)
2. Workflow Management Coalition: Process definition interface – XML process definition language version 2.1a. Technical Report WFMC-TC-1025, Workflow Management Coalition, Hingham, MA, USA (2008)
3. Workflow Management Coalition: XPDL implementations (June 2010), http://www.wfmc.org/xpdl-implementations.html (accessed May 21, 2010)
4. Rozenberg, G. (ed.): Handbook of Graph Grammars and Computing by Graph Transformation. Foundations, vol. I. World Scientific Publishing Co., Inc., River Edge (1997)
5. Jakumeit, E., Buchwald, S., Kroll, M.: GrGen.NET. International Journal on Software Tools for Technology Transfer, STTT (2010)

6. Heckel, R.: Tutorial introduction to graph transformation. In: Ehrig, H., Heckel, R., Rozenberg, G., Taentzer, G. (eds.) ICGT 2008. LNCS, vol. 5214, pp. 458–459. Springer, Heidelberg (2008)
7. Habel, A., Pennemann, K.h.: Correctness of high-level transformation systems relative to nested conditions† Mathematical. Structures in Comp. Sci. 19(2), 245–296 (2009)
8. Van Gorp, P., Mazanek, S., Rensink, A.: Transformation Tool Contest – Awards (2010), http://is.ieis.tue.nl/staff/pvgorp/events/TTC2010/?page=Awards
9. Van Gorp, P.: BPMN semantics: online virtual machine (2010), http://is.ieis.tue.nl/staff/pvgorp/share/?page=ConfigureNewSession&vdiID=364
10. Budinsky, F., Brodsky, S.A., Merks, E.: Eclipse Modeling Framework. Pearson Education, London (2003)
11. Wong, P.Y., Gibbons, J.: A process semantics for BPMN. In: Liu, S., Maibaum, T., Araki, K. (eds.) ICFEM 2008. LNCS, vol. 5256, pp. 355–374. Springer, Heidelberg (2008)
12. Wong, P.Y., Gibbons, J.: Formalisations and applications of BPMN. Science of Computer Programming (2009) (in Press, Corrected Proof)
13. Dijkman, R., Dumas, M., Ouyang, C.: Semantics and analysis of business process models in bpmn. Information and Software Technology (IST) 50(12), 1281–1294 (2008)
14. Prandi, D., Quaglia, P., Zannone, N.: Formal analysis of BPMN via a translation into COWS. In: Lea, D., Zavattaro, G. (eds.) COORDINATION 2008. LNCS, vol. 5052, pp. 249–263. Springer, Heidelberg (2008)
15. Raedts, I., Petkovic, M., Usenko, Y., van der Werf, J., Groote, J., Somers, L.: Transformation of BPMN models for behaviour analysis. In: Proceedings of the 5th International Workshop on Modelling, Simulation, Verification and Validation of Enterprise Information Systems, pp. 126–137. INSTICC Press (2007)
16. Dumas, M., Grosskopf, A., Hettel, T., Wynn, M.: Semantics of standard process models with or-joins. In: Meersman, R., Tari, Z. (eds.) OTM 2007, Part I. LNCS, vol. 4803, pp. 41–58. Springer, Heidelberg (2007)
17. Takemura, T.: Formal semantics and verification of BPMN transaction and compensation. In: Proceedings of the IEEE Asia-Pacific Conference on Services Computing, pp. 284–290. IEEE Computer Society, Los Alamitos (2008)
18. Roscoe, A.W.: The Theory and Practice of Concurrency. Prentice-Hall, Englewood Cliffs (1998)
19. van der Aalst, W.: Verification of workflow nets. In: Proceedings of the 18th International Conference on Application and Theory of Petri Nets, pp. 407–426 (1997)
20. Prandi, D., Quaglia, P.: Stochastic COWS. In: Krämer, B.J., Lin, K.-J., Narasimhan, P. (eds.) ICSOC 2007. LNCS, vol. 4749, pp. 245–256. Springer, Heidelberg (2007)

On a Study of Layout Aesthetics for Business Process Models Using BPMN

Philip Effinger[1], Nicole Jogsch[2], and Sandra Seiz[2]

[1] Wilhelm-Schickard-Institut für Informatik
[2] Wirtschaftswissenschaftliche Fakultät
Eberhard-Karls-Universität Tübingen, Germany
{philip.effinger,nicole.jogsch,sandra.seiz}@uni-tuebingen.de

Abstract. As BPMN spreads among a constantly growing user group, it is indispensable to analyze the expectations of users towards the appearance of a BPMN model. The user groups are mostly inhomogeneous since users stem from different backgrounds, e.g. IT, managerial sciences or economics. It is conceivable that BPMN novices may have different issues compared to higher skilled modeling experts. When this large set of users starts modeling, the expectations considering the graphical outcome of the modeling process may differ significantly.

In this work, we analyze layout preferences of user groups when modeling with BPMN. We present a set of layout criteria that are formalized and then confirmed by a user study. The conduction of the study reveals preferences of single user groups with respect to secondary notation and layout aesthetics. From our results, proposals for adaptions of software tools towards different BPMN users can be derived.

Keywords: user study, secondary notation, BPMN models, layout aesthetics.

1 Motivation

BPMN is a visual modeling language for business processes. The modeling toolbox comprehends elements such as activities, gateways, events etc. and connecting elements, e.g. associations, sequence or message flows. Also, BPMN provides pools and swimlanes which represent structural elements. When users begin to design a model they start on a blank sheet, or screen respectively, which implies that there are no hints for the user where to position the elements.

The development of an automatic layout algorithm that supports the user in this task is a mere logical progression. The layout algorithm could point to possible positions of new elements or could recalculate a 'good' layout from a given sketch. In [1,2], approaches for BPMN layout algorithms are undertaken. However, when developing a layout algorithm for a specific language as BPMN, one finds himself quickly in the state of asking: What actually matters a layout to be 'good'? What are the attributes that render a layout of a business process model to be readable and easy to grasp? What preferences or expectations towards a layout for business process models does a user have?

J. Mendling, M. Weidlich, and M. Weske (Eds.): BPMN 2010, LNBIP 67, pp. 31–45, 2010.
© Springer-Verlag Berlin Heidelberg 2010

From this starting point, we designed an empirical user study with the aim to reveal the attributes that actually matter for BPMN users when modeling and looking for 'good' layout of their diagrams. In this work, we refer to those attributes as layout aesthetics, as proposed in [3,4]. Our study also takes into account that modeling users of BPMN are an inhomogeneous set of people with diverse background and skills, e.g. education in process modeling or modeling experience in practice are important measurement factors [5].

From a cognitive point of view, layout aesthetics are known as *secondary notation* [5,6]. In contrast to *first notation*, in our case the graphical notation of BPMN, secondary notation carries extra information by other means than the official syntax [7]. According to [7], 'redundant recoding gives a separate and easier channel for information that is already present in the official syntax. Escape from formalism allows extra information to be added, not present in the official syntax'. As extra information in secondary notation, layout aesthetics are non-obvious cues in a graphical representation of a business process model. Although single aesthetics are not visible in a diagram but rather hidden in the layout, they are highly relevant for exploratory and modification activities [7] which both are very frequent tasks in the field of business process modeling.

In the user study, we support secondary notation in terms of layout aesthetics for BPMN. Moreover, we show preferences of specific user groups of modeling participants. We analyze layout aesthetics of users in comparison groups filtered by attributes 'gender', 'experience' and 'education'. The results can be used for designing user-supporting features in modeling tools and tools that contain layout algorithms and produce well-received layout for diagrams. The rest of the paper is organized as follows: In Section 2, we present our formal catalogue of layout aesthetics, followed by the summary of our test method in Section 3 and the presentation of our study conjectures and their evaluation in Section 4. Before we conclude, we give an overview on related research work in Section 5.

2   Layout Aesthetics

Layout aesthetics are a measure for a graphical property of a drawing. The goal of a layout algorithm is to fulfill given aesthetics or, in other words, to optimize the aesthetics requirements. For the development of an algorithm, layout aesthetics must be formalized and modeled in the algorithm. The formalizations must be known beforehand in order to adapt the layout algorithm accordingly.

For BPMN, results of research considering specific aesthetics are not known yet. However, for diagrams that are also graph-based, aesthetics were explored in studies, e.g. for diagrams in general [8] or specifically for UML-diagrams [9]. Since we consider BPMN to be a graph-based notation, the underlying structures correspond to graphs with BPMN elements as nodes and BPMN connection elements as edges. For general graphs, we can find aesthetics in the graph-drawing community that discusses general aesthetics. However, general graphs don't possess notation semantics, in contrast to BPMN where the graphical representation defines the type of the BPMN element, e.g. gateways have diamond shapes.

In [10], an overview on common graph drawing aesthetic criteria is given. Also, in [5], a subset of these criteria are confirmed to be important for BPMN.

The formalization of layout aesthetics is reached by expressing them as optimization problems, as in [3,4,11]. For BPMN as a graph-based notation, we consider the following layout aesthetics as necessary:

- Minimize the number of crossings of connecting elements (CROSSING).
- Minimize the area of the drawing (AREA).
- Minimize the number of bends of connecting elements (BEND).
- Minimize the number of overlapping (connection) elements (OVERLAP).
- Maximize the number of orthogonally drawn connecting objects (ORTHOGONAL).
- Maximize the number of connecting objects respecting workflow direction (FLOW).

Therefore, layout algorithms solve a multi-objective optimization problem. However, the task to tackle the whole set in a single algorithm is very complex which is underlined by the low number of tools supporting BPMN layout and the poor quality of layout results, see [1,12] for a quick overview.

Also, BPMN has specific requirements towards layout aesthetics since it provides notation semantics within its graphical representations. The following requirements represent aesthetics that consider the specific requirements of BPMN which can be derived inspecting the standardization document:

- Flow objects have different sizes (ELEMENT_SIZE).
- Partitions must be considered, e.g. pools and swimlanes (PARTITION).
- Labeling of pools, swimlanes and (connection) elements must be feasible (LABEL).

These principles are also mentioned in informal collections of drawing principles for special diagram types, e.g. for UML activity diagrams [13] or network diagrams [14]. Summing up the relevant aesthetics, we state the following list of standard layout aesthetics for the layout of BPMN diagrams:

CROSSING, ORTHOGONAL, AREA, BEND, PARTITION, LABEL, ELEMENT_SIZE, OVERLAP and FLOW.

The list of design standards is also conform on conventions within the BPMN community [1,15] and corresponds to a superset of supported aesthetics in existing BPMN tools.

In our test design, we developed a catalogue with statements that cover the different objectives of layout aesthetics. We face this catalogue with the users' personal appreciation of different layouts in order to confirm or reject our conjectures which will be proposed in Section 4.1.

3   Design of the User Study

3.1   Software Tools

In this Section, we will give a summary of our test design and test method. For a more detailed description, we refer to [12]. For the comparison of layout results for BPMN models, we have to choose among existing solutions of tools that support modeling with BPMN and provide automatic layout features. After a market study, we obtained a list with 54 software packages that support BPMN according to the vendors. However, support of BPMN only is not sufficient for our study, therefore, we developed the following criteria for tool selection:

1. BPMN support:
   The tool must support the standardized version 1.2 of BPMN, including all elements given in the standard.
2. Automatic layout support:
   The tool must provide the user with a feature that calculates an automatic layout of a given BPMN model. Thus, the user has the possibility to acquaint himself with the principles of layout aesthetics. The layout feature must also consider basic BPMN aesthetics, e.g. swimlanes must be respected and edges must be drawn orthogonally.
3. Evaluation license availability:
   For our study, we depend on evaluation licenses provided by the software vendors.

These criteria were compulsory. We could not consider tools or vendors that could not meet one or more of the criteria. During the study preparation, the most challenging criteria was the support of automatic layout. Also, software that supposedly supports BPMN provided repeatedly only a subset of BPMN. This prevented us from preparing our modeling test case on those tools.

After applying our criteria filters on the list of software tools, we obtained a set of five tools (three commercial tools plus two versions of a self-developed tool [1]) that fulfilled all criteria and that were considered for our user study. In [12], we provide a complete list with all software vendors that were considered.

3.2   Test Method

In the following, we will shortly describe the procedure instructions given to probands and the surrounding setup for the experiment of the study. The setup was successfully confirmed to be well-chosen with the help of a pre-test prior to the experiment.

The probands of the study were chosen among students with majors in economics and/or computer science. Their skills (education and/or experience) in process modeling spread from very low to very high in order to represent inhomogeneous but, because of their major subject, potential future users of modeling tools.

The software tools were installed on PCs and probands were randomly assigned to one tool. For assuring a basic common understanding on BPMN and business process modeling, probands initially were asked to insert a simple extension into an existing process using the assigned modeling tool. The set of BPMN elements needed for the implementation of the extension was given as support. The process for the modeling task was a book order instance. The process can be inspected in detail in Figure 6. After the modeling part, probands were requested to prepare a presentable version of the new process. They were advised to use the automatic layout feature(s) of the corresponding tool. At this point, probands were supposed to be familiar with BPMN modeling and also with the automatic layout results and features of the respective tool. For the evaluation, each proband obtained five different layout versions of the new process, one from each evaluated tool. In the accompanying questionnaire, probands were asked to give rankings among the tools for each statement contained in the questionnaire. For comparison reasons, we chose to ask for rankings with strictly relative order among the tools. Thus, two tools cannot be ranked equally and we obtain more exact responses. The statements of the questionnaire correspond to the layout aesthetics of Section 2. All layout aesthetics are represented by a statement. For the representation as statements, we group the statements into categories that aimed at the same attribute of a graph/diagram. The following list gives the wording of the statements as they appeared in the questionnaire:

1. Category 1: "Connection elements" (edges)
    1.1 Edges are drawn orthogonally and are inserted in such a way that they appear as short as possible (ORTHOGONAL).
    1.2 Edges appear to be drawn with the lowest possible number of crossings (CROSSING).
    1.3 Edges appear to be drawn with the lowest possible number of bends (BEND).
    1.4 Edges are drawn such that they consider the reading direction (FLOW).
2. Category 2: "Area usage"
    2.1 The size of the swimlanes is chosen such that all elements have enough space (ELEMENT_SIZE).
    2.2 The diagram contains unused space that could be better exploited by rearranging the elements (AREA).
3. Category 3: "Elements"
    3.1 Elements are arranged such that they do not overlap (OVERLAP).
    3.2 The size of the elements is chosen such that the description of the label is readable (LABEL).
    3.3 The assignment of an element to its swimlane is easy to perceive (PARTITION).
4. Category 4: "Coloring"
    4.1 The choice of colors supports to obtain a detailed comprehension of the model.
    4.2 The choice of colors supports to obtain a quick overview of the model.

After the ranking of tools for each statement, probands were asked to rank the categories in decreasing importance according to their personal judgement. Also, probands were encouraged to add additional categories and/or statements that to their opinion were not represented in our above catalogue of statements. These proband-defined categories were also included in the ranking of categories. Eventually, the probands are asked to give a ranking of all layout diagrams presented in the file to their personal appearance preference, independently from the above given statements.

The total time per proband of our study depended on the skill level of the probands. The periods of time ranged from 55-75 minutes per proband for the task of modeling and responses to the two questionnaires that considered layout and usability questions. The total number of participants of the study was 39.

4   Evaluation

In this section, we present the results of our study. The study data was collected while conducting the study at two universities, at Eberhard Karls Universität Tübingen and Humbold Universität Berlin. The circumstances were kept comparable during both times, identical software tools and identical questionnaires were used.

Note that generally on ordinal scales, the application of mean values is not possible. However, in our case, as stated in [16], the distance between the ordinal items can be considered to be equal from a proband's perspective. Therefore, the application of mean values for our evaluation is rendered possible and allows an thorough evaluation using descriptive statistics.

First, we analyze the categories of Section 3.2. For the categories, we present the ranking of all users. For the analysis, we also examine the results of the proband group separately by gender, process modeling experience and prior education in modeling.

For the analysis, the total user group is filtered yielding to subset groups with the following attributes:

- Gender: male (m) and female (f)
- Experience: business process modeling experience is rated from 'none' (0) and 'low' (1) to 'high/very high' (2+).
- Education: number of lectures (or similar events) attended in the field of business process modeling; the group is split into ranges from none (0), one (1) to two or more (2+).

The division into subset groups is due to the focus on inhomogeneous user groups. Each subset group represents a set of business process modelers with distinct backgrounds, capabilities and skills. Thus, we analyze the corresponding preferences of these groups. Before that, we propose our conjectures for the analysis in the following section.

4.1   Conjectures

Our conjectures aim to confirm that the set of statements that we defined in Section 3.2 is convenient to support the aesthetics from Section 2 and correspond to the users' preferences.

For better coverage, we state our conjectures for the total group and the subset groups separately. The conjectures will also be evaluated separately in Section 4.3.

Conjecture 1 (Layout Aesthetics for Total Group). The user rating of the total group with respect to 1) the catalogue of category statements and 2) the users' personal appreciation (general ranking) correspond.

If Conjecture 1 is supported, we can state that the statements of our catalogue of categories 1-4 are sufficient to fulfill the aesthetics' requirements of a BPMN diagram for all probands.

Conjecture 2 (Layout Aesthetics for Subset Group 'Gender'). The user rating of the subset groups 'Gender' (male/female) with respect to 1) the catalogue of category statements and 2) the users' personal appreciation (general ranking) correspond.

If Conjecture 2 is supported, we can confirm that the statements of our catalogue of categories 1-4 are sufficient to fulfill the aesthetics' requirements for the subset group of male, or female respectively, probands.

Conjecture 3 (Layout Aesthetics for Subset Group 'Experience'). The user rating of the subset groups 'Experience' (Experience (0), Experience (1), Experience (2+)) with respect to 1) the catalogue of category statements and 2) the users' personal appreciation (general ranking) correspond.

Conjecture 3 states the validity of the statements considering probands with different levels of modeling experience, from none (novices) to very high (experts).

Conjecture 4 (Layout Aesthetics for Subset Group 'Education'). The user rating of the subset groups 'Education' (Education (0), Education (1), Education (2+)) with respect to 1) the catalogue of category statements and 2) the users' personal appreciation (general ranking) correspond.

If Conjecture 4 is supported, we can confirm that the statements are sufficient to fulfill the aesthetics' requirements for the subset groups of probands with different educational background for business process modeling.

If the correspondence between the general ranking and the catalogue of statements is not given, that means that Conjectures 1- 4 cannot be supported, we state a conjecture for the case that, from the resulting values, we can derive a tendency to the better- and the less well-rated tools.

Conjecture 5 (Layout Aesthetics and Tool Tendency). From the user ratings of the Total Group/Subset groups 'Experience'/'Education'/'Gender', we can observe a match between the best- and the worst-rated tools with respect to 1) the catalogue of category statements and 2) the users' personal appreciation (general ranking).

In addition to conjectures on whole subset groups, we state conjectures that claim the statements of our layout aesthetics adapt better the higher the users' experience or the better their education in business process modeling with BPMN. In other words, the differences between the general ranking and the catalogue of aesthetics' statements diminish inside the subset group 'Education', or 'Experience' respectively:

Conjecture 6 (Increasing Benefit with Higher Experience). The differences in the user rating of the subset groups 'Experience' with respect to 1) the catalogue of category statements and 2) the users' personal appreciation (general ranking) diminish with increasing experience levels of probands.

Conjecture 7 (Increasing Benefit with Higher Education). The differences in the user rating of the subset groups 'Education' with respect to 1) the catalogue of category statements and 2) the users' personal appreciation (general ranking) diminish with increasing educational background of probands.

Conjectures 6 and 7 represent our conjecture that our layout aesthetics are more efficient and exceedingly appropriate for BPMN modeling users that have a fundamental understanding of process modeling in general. Before analyzing our set of conjectures, we present the results of the ranking of categories.

4.2   Results of Category Ranking

The results of the category ranking show where the subset groups see the aesthetic with the largest impact on the layout. The categories were ranked in a relative order from 1 to 4, respectively to 5 or 6 if proband-defined categories were given. The diagram in Figure 1 depicts the resulting values. After normalization, values can range from 0 to 4 while 4 is the highest (best) score. The following interpretation of Figure 1 is done for each category.

For category 1 "Connection elements", the values contain only two spikes, for group 'Experience (1)' and group 'Experience (2+)'. However, the derivations of the spikes are diametrically opposed. Thus, no conclusion can be drawn from these spikes. However, together with the highest mean value of all categories, the low number of derivations show that category 1 has a great effect on the acceptance of the users. Thus, we can state that users prefer layout models that correspond with statements 1.1 - 1.4.

Category 2 "Area usage" has the largest spikes in the 'Education' subset group, more precisely it has an obvious peak in 'Education (0)'. This peak renders category 2 to be the most important for 'Education (0)', in contrast to other subset groups. For subset groups 'Education (1)' and 'Education (2+)',

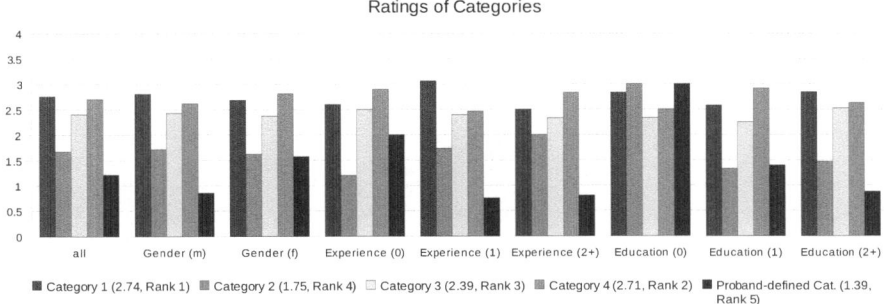

Fig. 1. Ratings of the categories for total group (all) and all subset groups. Mean values and ranks for categories are given in parentheses.

the derivations are low. Thus, category 2 seems to lead to a most promising but eventually alleged effect on the layout for users with less education in process modeling. In total, category 2 ranks 4th and therefore last of all pre-defined categories.

Category 3 "Elements" is the category with the significantly least derivations in all subset groups. Thus, all users agree that element layout properties are indispensable, but no group rated this category 1st, in other words most preferred. Rank 3 for this category shows that it contributes important statements and may not be ignored when creating layout models.

Category 4 "Coloring" has only two minor spikes. In subset groups 'Experience (1)' and 'Education (1)', the spikes show two derivations to slightly higher, or lower respectively, values. However, in most subset groups, category 4 is rated most important or second most important after category 1. Thus, together with category 1, this category has strong effect on the layout.

The proband-defined category shows the most diverging behaviour. While for subset group 'Education (0)', the category yields the highest value, it ranks last for 6 of 9 subset groups. Only subset group 'Experience (0)' describes a similar behaviour as 'Education (0)'. The value of 'Experience' and 'Education' may already imply that the more experience or education modeling users are equipped with, the more the rating of models converges to a set of commonly accepted measurement attributes and aesthetic criteria (Conjectures 6 and 7).

4.3   Overall Ranking

We will now face the results of the general ranking given by the probands with the aggregation over the results of the set of categories. Therefore, we calculate the mean values of the general ranking for both, the total group and each subset groups. This approach allows to consider the subset groups and their inhomogeneities. For obtaining the results, we compare data from the general ranking (GR) over all tool models with the aggregated data per category and its corresponding statements from Section 3.2 (mean of categories, MC1-4).

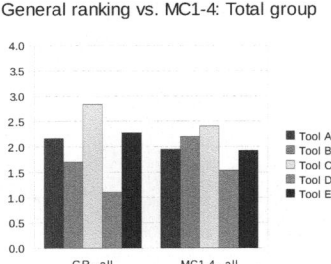

Fig. 2. Comparison of general ranking (GR) and mean of categories (MC1-4) for total group

By Total Group:

The contrast between the GR and MC1-4 is depicted in Figure 2 and becomes visible when taking a closer look at Tools A,B and E. Their ranking is not equal over both measurements. However, as Tool C always ranks 1st and Tool D ranks last, we can state that our methods can distinct between tools that suffice users' preferences and those that fail. Thus, we have to reject Conjecture 1, but the support of Conjecture 5 for the total group is successful.

By Gender:

The gender subset group contains the male and female subsets of probands. For the comparison between GR and MC1-4, depicted in Figure 3, we state that a link between GR and MC1-4 is not given. The ratings of both, male and female, differ for GR and MC1-4. However, in all subset groups, Tool C ranks 1st and Tool D ranks last which, as in the total group, allows us to distinct between best and imperfect tools. Therefore, Conjecture 2 must be rejected, but Conjecture 5 is supported for subset group 'Gender'.

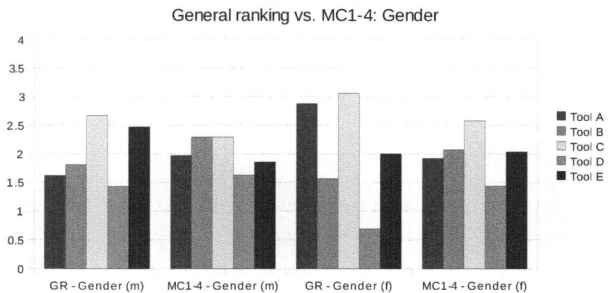

Fig. 3. Comparison of general ranking (GR) and mean of categories (MC1-4) for subset groups of 'Gender'

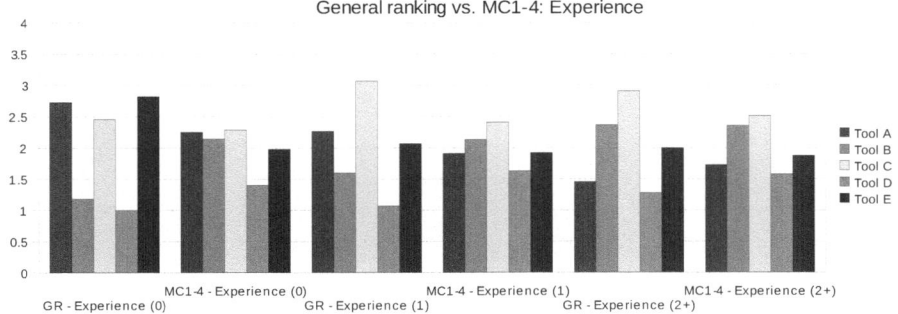

Fig. 4. Comparison of general ranking (GR) and mean of categories (MC1-4) for subset groups of 'Experience'

By Experience:
The subset groups for filter 'Experience' sum up to 6 groups, see Figure 4. For 'Experience (0)', the values and rankings of GR and MC1-4 differ, but considering users with at least little experience in group 'Experience (1)', we observe only a minor switch of ranks for Tool A and B. For users with high experience in group 'Experience (2+)', the values correspond and the ranking is identical.

Since the values of 'Experience (0)' do not correspond, we have to reject Conjecture 3. However, since the differences clearly diminish with higher experience of modeling users, Conjecture 6 is supported. Also, for both subset groups 'Experience (1)' and 'Experience (2+), we can confirm that Conjecture 5 is supported since one can obviously inspect the best and the deficient tools chosen by the probands.

By Education:
The filter 'Education' also creates 6 subset groups. In Figure 5, the values are presented. We can state that for 'Education (0)', the values diverge, there are

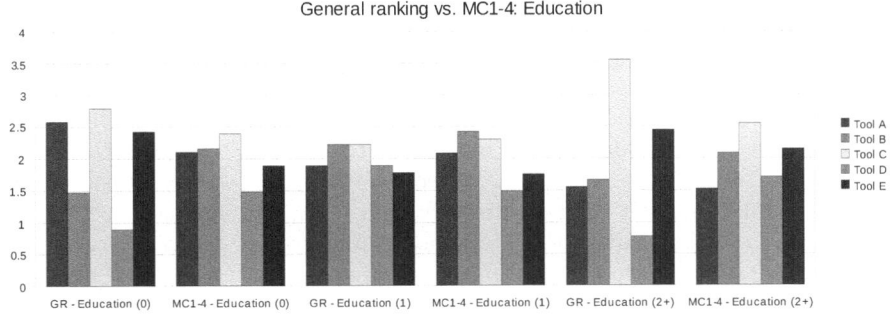

Fig. 5. Comparison of general ranking (GR) and mean of categories (MC1-4) for subset groups of education

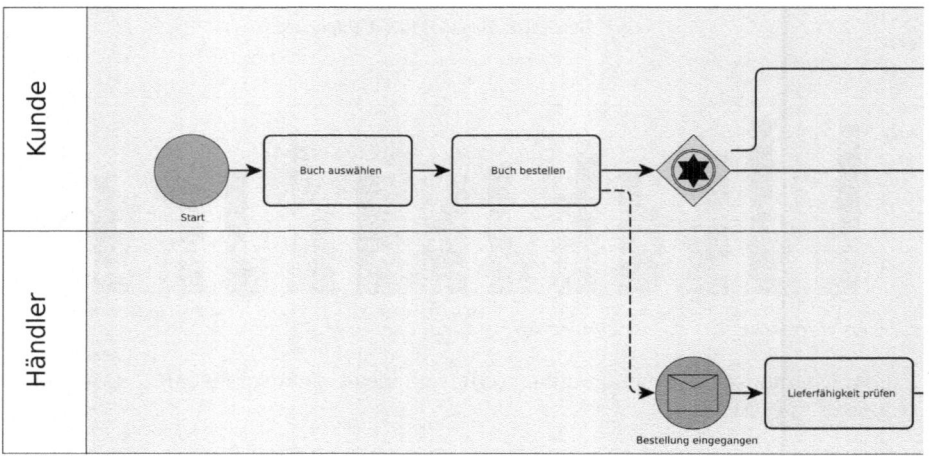

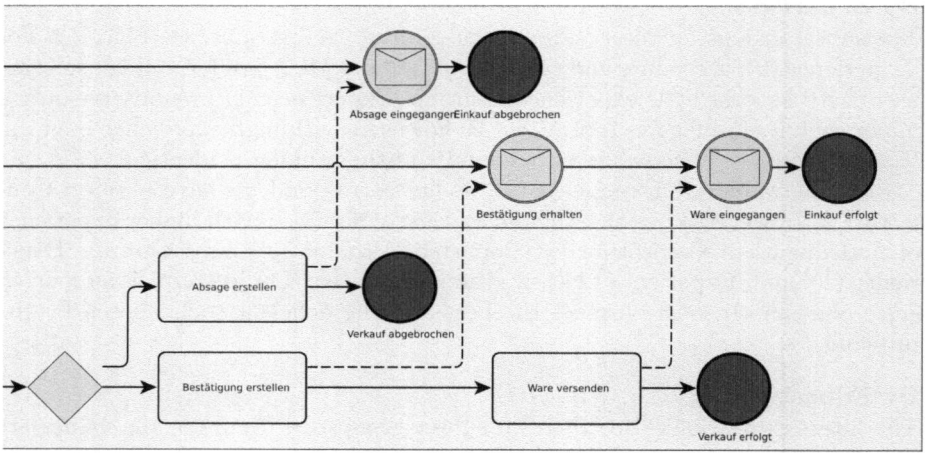

Fig. 6. Layout of Tool C scored highest in 7 of total 9 subset groups. For reading purposes, the image is for this paper manually cut in two halves of equal width; the upper first part is originally directly connected to the lower second part. As the study was conducted in German, the labels are given in German language.

two swaps of tools' ranking (Tool A,B and E). However, for groups with higher modeling education, we only notice one swap: Tool D and E for 'Education (1)', Tool A and D for 'Education (2+)'. Moreover, the distance between values that cause swaps diminish the higher the education level, e.g. the distance between Tool A and D for 'Education (2+) is almost vanished to 0.19 on a 0-to-4-scale.

Also, as for subset group 'Experience', we cannot state that the corresponding Conjecture 4 for group 'Education' is supported. On the other hand, we can confirm an improvement of the results the higher the educational background, therefore Conjecture 7 is supported. Moreover, Conjecture 5 is supported for 'Education (0)' and clearly for 'Education (2+), too.

Before closing the evaluation section, we sum up the results and conjectures of this section. Conjectures 1, 2, 3 and 4 are rejected since the differences between the values of the general ranking (GR) and the mean of categories (MC1-4) diverge in the total group and at least one subset group of the filters (gender, education, experience). However, we confirm Conjecture 5 for 7 of total 9 subset groups. The Conjecture states the correct prediction of the user preference tendency of a tool's layout capabilities when using our catalogue of statements. Also, Conjectures 6 and 7 can be confirmed. They state that the values of subset groups 'Experience' and 'Education' correspond for GR and MC1-4 when considering the groups with average or higher practice experience or at least basic education respectively for the field of business process modeling.

The layout from Tool C that was favoured by the probands in the general ranking is depicted in Figure 6.

5   Related Work

The term of secondary notation is due to fundamental cognitive research of [6,17]. Secondary notation is an important part of the cognitive dimensions of notation framework developed by [7]. It is also used in the study design of [5] where differences in understandability of business process models between novices and experts are targeted. In [18], graph aesthetics are used for cognitive measurements.

The area of research on layout aesthetics is broader if considering other diagram types or graph classes, too. There are diagram types widely related to BPMN, e.g. UML-diagrams which were part of an analysis in [19] where laws for diagram layout were formalized, e.g. the 'law of proximity'. This would refer to our statements 'ORTHOGONAL' and 'ELEMENT_SIZE'. Aesthetics for UML-diagrams were also proposed in the context of automatic layout in [20]. Suggestions of aesthetics are also given for Petri-Nets in [21] or for more general graphs in [22]. In [23], metrics are proposed and validated for entity-relationship (ER-) diagrams. Aesthetics, or 'effects', for social-network visualization was considered in [14] with the focus on edge crossings, in our case 'CROSSING'.

The set of layout aesthetics is further enriched by research works and studies of [8,9,24] where a subset of aesthetics is target of an ranking analysis. Further confirmation for aesthetics ranking is conducted by [25] that states that diagrams with short edges can be read more easily because the nodes' proximity is higher and the probability of crossings ('CROSSING') is lower. Also, [26] confirmed the importance of 'CROSSING'. Moreover, general modeling guidelines, e.g. [25], are available and provide a fundamental set of aesthetics.

User studies concerning aspects of layout aesthetics are done by [9] for UML, [15] for syntactical structures in process models and also [27] for general graphs. In [5], the focus is on secondary notation. The cognitive complexity of integrating multiple diagrams with different notations is examined in [28].

6   Conclusion

In this work, we analyzed secondary notation in terms of layout aesthetics and users' preferences of layout aesthetics for BPMN with consideration of inhomogeneous user groups. We proposed a catalogue of criteria that promise modeling results that are well-accepted by most users when being applied.

The formalized catalogue is tested by the conduction of a user study. The results of the study were presented and interpreted. The data analysis of the study results was performed with respect to not only all participants at a time but also to subset groups according to modeling experience, modeling education and gender. We were able to show that our layout catalogue is most appropriate when applied for users with average or higher practice experience and users with at least basic education in business process modeling. We also could show that our catalogue of statements is sufficient to predict the tendency of the users' judgement for the layout capabilities of a tool. For future works, we like to address a broader set of probands, not only in quantity but also in consideration of more diverse backgrounds, e.g. culture and/or nationality.

Concluding, our results can be used for designing powerful algorithms in modeling tools for BPMN that produce layout for BPMN diagrams that will be well-received by users.

References

1. Effinger, P., Siebenhaller, M., Kaufmann, M.: An Interactive Layout Tool for BPMN. IEEE International Conference on E-Commerce Technology 1, 399–406 (2009)
2. Kitzmann, I., König, C., Lübke, D., Singer, L.: A Simple Algorithm for Automatic Layout of BPMN Processes. In: CEC, pp. 391–398 (2009)
3. Siebenhaller, M., Kaufmann, M.: Drawing activity diagrams. Technical Report WSI-2006-02, Wilhelm-Schickard-Institut (2006)
4. Siebenhaller, M., Kaufmann, M.: Drawing activity diagrams. In: Proceedings of ACM 2006 Symposium on Software Visualization, SoftVis 2006, pp. 159–160. ACM, New York (2006)
5. Schrepfer, M., Wolf, J., Mendling, J., Reijers, H.A.: The impact of secondary notation on process model understanding. In: PoEM, pp. 161–175 (2009)
6. Petre, M.: Why looking isn't always seeing: Readership skills and graphical programming. ACM Commun. 38(6), 33–44 (1995)
7. Green, T.R., Blackwell, A.F.: A tutorial on cognitive dimensions(1998) (last accessed 2010-05-31)
8. Purchase, H.C.: Which aesthetic has the greatest effect on human understanding. In: DiBattista, G. (ed.) GD 1997. LNCS, vol. 1353, pp. 248–261. Springer, Heidelberg (1997)
9. Purchase, H.C., Allder, J.A., Carrington, D.A.: User preference of graph layout aesthetics: A UML study. In: Marks, J. (ed.) GD 2000. LNCS, vol. 1984, pp. 5–18. Springer, Heidelberg (2001)
10. Tamassia, R., DiBattista, G., Eades, P., Tollis, I.: Graph Drawing. Prentice Hall, Englewood Cliffs (1999)

11. Siebenhaller, M.: Orthogonal Drawings with Constraints: Algorithms And Applications. PhD thesis, Wilhelm-Schickard-Institut, University of Tuebingen (2009) (to appear)

12. Seiz, S., Effinger, P., Jogsch, N., Wehrstein, T.: Forschungsprojekt: Usability-Evaluation von BPMN-konformer Geschäftsprozessmodellierungssoftware. In: Arbeitsberichte zur Wirtschaftsinformatik 35, Lehrstuhl für Wirtschaftsinformatik, Universität Tübingen (April 2010) (German)

13. Ambler, S.W.: The Elements of UML 2.0 Style. Cambridge University Press, Cambridge (2005)

14. Huang, W., Hong, S.H., Eades, P.: Effects of sociogram drawing conventions and edge crossings in social network visualization. J. Graph Algorithms Appl. 11(2), 397–429 (2007)

15. Mendling, J., Reijers, H.A., Cardoso, J.: What makes process models understandable? In: Alonso, G., Dadam, P., Rosemann, M. (eds.) BPM 2007. LNCS, vol. 4714, pp. 48–63. Springer, Heidelberg (2007)

16. Gehring, U.W., Weins, C.: Grundkurs Statistik für Politologen und Soziologen, 5th edn. VS Verlag für Sozialwissenschaften (2009) (German)

17. Petre, M.: Cognitive dimensions 'beyond the notation'. J. Vis. Lang. Comput. 17(4), 292–301 (2006)

18. Ware, C., Purchase, H.C., Colpoys, L., McGill, M.: Cognitive measurements of graph aesthetics. Information Visualization 1(2), 103–110 (2002)

19. Sun, D., Wong, K.: On evaluating the layout of UML class diagrams for program comprehension. In: IWPC, pp. 317–326 (2005)

20. Eichelberger, H.: Aesthetics and Automatic Layout of UML Class Diagrams, PhD thesis, Universität Würzburg (2005)

21. Jensen, K.: Coloured Petri nets: basic concepts, analysis methods and practical use. Monographs in Theoretical Computer Science. An EATCS Series, vol. 2. Springer, Heidelberg (1996)

22. Coleman, M.K., Parker, D.S.: Aesthetics-based graph layout for human consumption. Software – Practice and Experience 26(12), 1415–1438 (1996)

23. Genero, M., Poels, G., Piattini, M.: Defining and validating metrics for assessing the understandability of entity-relationship diagrams. Data Knowl. Eng. 64(3), 534–557 (2008)

24. Purchase, H.C., Cohen, R.F., James, M.I.: Validating graph drawing aesthetics. In: Brandenburg, F.J. (ed.) GD 1995. LNCS, vol. 1027, pp. 435–446. Springer, Heidelberg (1996)

25. Apfelbacher, R., Knöpfel, A., Aschenbrenner, P., Preetz, S.: FMC visualization guidelines (2006), http://www.fmc-modeling.org/visualization_guidelines

26. Huang, W., Hong, S.H., Eades, P.: Effects of crossing angles. In: PacificVis, pp. 41–46 (2008)

27. Huang, W., Eades, P., Hong, S.H.: Beyond time and error: a cognitive approach to the evaluation of graph drawings. In: BELIV 2008: Proceedings of the 2008 Conference on BEyond Time and Errors, pp. 1–8. ACM, New York (2008)

28. Hahn, J., Kim, J.: Why are some diagrams easier to work with? effects of diagrammatic representation on the cognitive intergration process of systems analysis and design. ACM Trans. Comput.-Hum. Interact. 6(3), 181–213 (1999)

The Role of BPMN in a Modeling Methodology for Dynamic Process Solutions

Jana Koehler

IBM Research - Zurich, 8803 Rüschlikon, Switzerland

Abstract. This paper introduces a design method for dynamic business process management solutions in which the well-known modeling elements of *business object life cycles*, *business rules*, and *business activities* are integrated in a distributed system as equal communicating components. Using the EURENT car rental domain originally developed by the business rules community, it is demonstrated how this method can be used to enable adhoc and rule-driven activities integrated with the life cycle management of business objects. A modeling methodology based on BPMN collaboration diagrams is proposed to describe component interactions and behavior. Agile principles are applicable to incrementally build the solution in which scenarios play a major role to validate and further evolve the solution's behavior. A clear separation between components, their interaction, and details of the internal component behavior facilitates change and the implementation of business patterns.

1 Towards Dynamic Process Solutions

Over the past years, the Business Process Management (BPM) community has been increasingly discussing more flexible BPM solutions that have a stronger focus on data and offer simple, yet powerful ways to integrate business rules. For example, the authors of [1] discuss case handling as a paradigm for dynamic BPM solutions and identify the following requirements:

- Avoid context tunneling and provide all information relevant to a case as needed, not just the information subset required to execute an activity within a predefined process flow.
- Decide which activities are enabled on the basis of the available information rather than enforcing a specific activity flow.
- Separate work distribution from authorization instead of using activity-based routing as a single mechanism to simultaneously address both requirements.

In addition, a dynamic BPM solution must support actor-initiated activities, which means that the human participants in a business process should be able to decide when to perform a business function. However, business conduct as captured in business rules should always be followed and monitored. Business rules should thus not only be used as automated decisions as it is common today, but they should primarily serve to establish obligations for actors to perform activities within the boundaries of appropriate business conduct.

J. Mendling, M. Weidlich, and M. Weske (Eds.): BPMN 2010, LNBIP 67, pp. 46–62, 2010.

This paper proposes a modeling methodology that addresses the above-mentioned requirements. It proposes a shift from the explicit modeling of predefined end-to-end processes to an agile design approach where the business capabilities (*activities*) of actors, the *business rules* that initiate and govern actor behavior, and the *life cycles* of the main *business objects* move into the primary focus of attention.

Activities, rules, and life cycles are considered as three types of equal communicating components in a distributed system. None of the three component types is subordinate to the other; they interact via message exchanges with each component type having different capabilities. Whereas activities, business rules, and object life cycles are well-known modeling elements, their equal interaction and uniform representation has not been investigated so far. The paper explores the consequences of such an *integrated approach* for a BPM solution.

Instead of proposing a new modeling language, the approach relies on BPMN 2.0 [2] and introduces a modeling methodology using BPMN collaboration diagrams to specify a dynamic process solution. By using BPMN, a uniform representation of activity, rule, and object life cycle components is achieved, which integrates all functional aspects of a BPM solution in a seamless manner. As BPMN 2.0 strives for a new level of integrating business-user-friendly modeling with direct model execution, the immediate exploration and simulation of process scenarios during modeling will help to build dynamic process solutions of high quality.

The paper uses a subset of the EURENT domain that models the typical operations of the branch office of a car rental company containing many adhoc activities and more than 100 business rules [3,4]. To validate the proposed methodology, parts of the domain were implemented using the process engine ePVM [5]. As the methodology is independent of a particular runtime engine, the discussion in this paper focuses on the conceptual pillars and abstracts from implementation details.

The paper is organized as follows: Section 2 introduces the main idea and shows examples of the three component types business rules, object life cycles, and actors in the EURENT domain. In Section 3, a 3-layer architecture as the foundation of the dynamic BPM solution is discussed, whereas Section 4 introduces principles of component interaction and BPMN collaboration diagrams as a means to specify component communication and behavior. Section 5 identifies next steps in research to further mature the proposed approach. Section 6 summarizes related work and Section 7 concludes.

2 Actors, Rules, and Object Life Cycles as Equal Communicating Components

The basic idea of the proposed modeling methodology is simple. Instead of modeling a process by capturing the flow of activities, refining this flow with data, adding roles to activities, and eventually refining decisions with business rules, the focus shifts to actors and their business capabilities, business objects capturing the main data and possessing a life cycle, as well as business rules establishing obligations for actors. Actors, object life cycles, and business rules form three types of communicating components where none has control over the other. The interaction of the components using messages creates the end-to-end process.

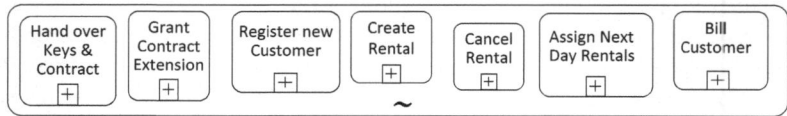

Fig. 1. Capabilities of the front desk clerk actor as activities in a BPMN adhoc process

Actors have certain capabilities and perform business functions by acting in a particular role. These capabilities are modeled as BPMN activities. For example, in the EURENT domain, three main roles possessing numerous activities can be distinguished:

– The **branch manager**, who oversees a particular branch office of the rental car company, is responsible to *buy*, *sell*, or *transfer cars*, *keep the contact to the police* in case of missed cars, and *bar customers* who misbehaved when renting a car.
– The **car park assistant** oversees all technical operations on the car fleet, including activities to *service cars* and *receive returned cars*.
– The **front desk clerk** interacts with the customer. For example, she will *hand over the keys and the contract* when the customer arrives to pick up the car, she might *grant a contract extension* upon a call of a customer, which may well interrupt the *registration of a new customer* who waits at the counter.

Figure 1 shows some of the activities of the front desk clerk in a BPMN adhoc process. It is obvious that it would not make sense to establish a predefined control-flow between these activities as it could never capture all reasonable process instances.

Business objects represent the main information objects of a business process. Identifying business objects is an established modeling step when designing a BPM solution. The EURENT domain contains very detailed data models from which the main business objects can be extracted, such as the *rental contract*, the *customer record*, and the *rental car*. Business rules also provide important information about business objects. For example, the structural rule *"a Customer has at least one of the following: a Rental Reservation, an in-progress Rental, a Rental completed in the past 5 years"* provides further insight into the concept of the customer record and suggests possible subtypes or states of rental contracts that should be distinguished.

The methodology discussed in this paper associates a state machine with each main business object that represents its life cycle. The methodology proposes that the object life cycle controls any create, read, delete, and update (CRUD) operations on the business objects as it is common in object-oriented programming encapsulation. The encapsulation is extended by making the life cycle explicit as a communicating component, which implies that business objects cannot be claimed as resources under possession by an actor, who would then be sure that executing an activity changes the state of the object. Instead, an actor can only send a message to an object requesting the object to change state and has to implement its own error-handling behavior in case the request is not successful. Business logic for the manipulation of business objects is decomposed along valid state transitions of the life cycle and represented in BPMN collaboration diagrams, see Section 4.

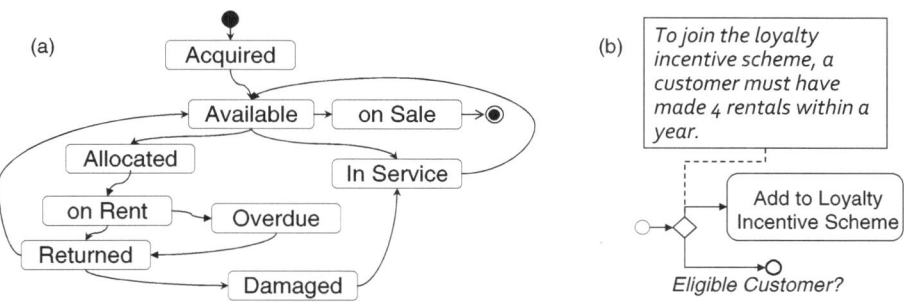

Fig. 2. Lifecycle of a rental car and example of a business rule automating a decision in a process

State information can be extracted from the EURENT domain models, but it is also worth noting that business rules are an important source for further detailing a life cycle. For example, the rules *"a car from another branch may be allocated, if there is a suitable car available and there is time to transfer it to the pick-up branch"* and *"if a car is three days overdue and the customer has not arranged an extension, insurance cover lapses and the police must be informed"* reveal three important states of the car business object. A possible life cycle for the car business object is shown in Figure 2(a).

More than 100 business rules were extracted from the EURENT domain documents that can be grouped into *structural* and *operative* rules. Structural rules capture necessary characteristics of business objects and are primarily used to ensure data integrity throughout the business processes. An example is shown in Figure 2(b) where the rule defines the necessary condition for a customer to be eligible for a bonus program. Operative rules formulate obligations and govern the conduct of business activity. They focus directly on the propriety of conduct in circumstances where willful or uninformed business activity can fall outside the boundaries of behavior deemed acceptable [3] and are thus of particular importance to the methodology proposed in this paper. The proposed methodology is independent of a particular rules language, but encapsulates rules into communicating components, which allows a rule to receive events of interest and initiate activities or changes on business objects by sending messages to actors and business object life cycles. Consequently, business rules are no longer passive structures evaluated or manipulated by business processes, see Section 4 for details.

3   Architecture and Influence of the Communication Infrastructure

The modeling methodology is associated with a 3-layer architecture that reuses proven principles from service-oriented architectures (SOA). This architecture was developed for the implementation of the EURENT dynamic BPM solution in ePVM. It is shown in Figure 3. In this architecture, the first layer comprises the presentation layer, which is clearly separated from the business logic captured in the other layers. Although the view of the actors on the business objects and other related information (for example business rules and other actors) is very important, it should not be mixed with the business logic.

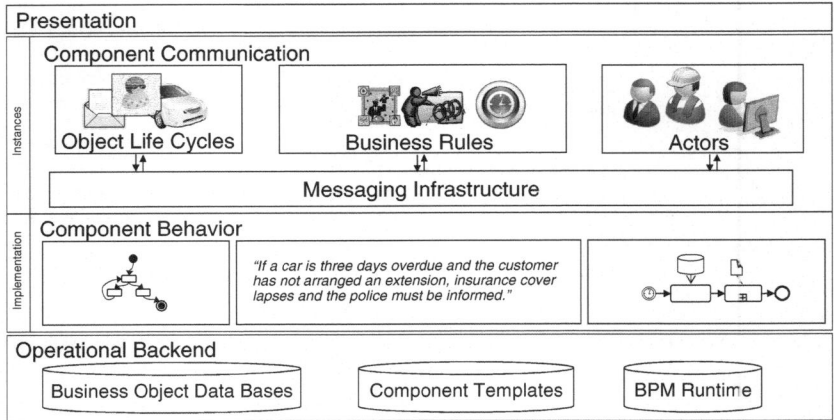

Fig. 3. Layered architecture with actors, object life cycles, and business rules as equal communicating components

In contrast, forms appeared as an almost integral part of the definition of business activities in [1]. The design and integration of the presentation layer can exploit common technology and is outside the scope of this paper as are also technical details that would address non-functional requirements such as scalability and performance.

The second layer focuses on the communication of the component instances separated from the definition/implementation of the internal component behavior. Business rules, object life cycles, and actors are modeled as communicating components that rely on a common messaging infrastructure. The proposed methodology studies a setup in which all components act as equal partners, i.e., none has control over the other and they only interact through messages. The reaction to a particular message is under the control of a component. At this layer, the message interfaces of each component are defined and principles of interaction are established, clearly separated from details of the internal component behavior. In addition, it is assumed that monitoring components (or notification broadcasters) are provided by the messaging infrastructure to help the components communicate with each other. BPMN collaboration models are used to describe the component communication as well as the internal component behavior, see Section 4 for further details.

The third layer represents the operative backend systems: the runtime engine used, the data bases in which business objects and templates for the object life cycles, business rules, and activities are stored. For example, the runtime provides functionality to create new instances of business object life cycles, activities, and rules based on the available model templates. Data bases hold information about actors, business objects, and rules, and can be queried by the components.

The messaging infrastructure influences the concrete design of the components. The subsequent sections assume that components can directly send messages to each other and also use a publish/subscribe mechanism, possibly moderated by monitors, to receive information of interest published by other components. Such a setup is common in today's business infrastructures. However, the proposed methodology can also be tailored to a different communication infrastructure, with the component communication changing accordingly.

4 Component Communication and Interaction Principles

This section establishes basic principles for the interaction behavior of the components and introduces BPMN collaboration diagrams for each component type. A component follows a processing cycle where it

1. receives an incoming message that acts as the primary trigger to initiate some behavior of the component,
2. sends and receives further secondary messages during the course of the behavior to determine its reaction to the primary trigger,
3. decides on the reaction, which can result in a reply to the originator of the primary trigger,
4. terminates its life or waits for the next primary trigger message.

Figure 4 summarizes the processing cycle that is inspired and natively supported by the ePVM runtime. It also corresponds to the conventional activation model of business processes that are instantiated by some start event.

Business rules and business object life cycles always require an explicit primary trigger message from another component. Business activities can be directly executed by an actor without any foregoing primary trigger message, because actors often self-initiate a processing cycle as a reaction to an externally observed event that happens in their environment. For example, the manager of an airport branch decides to buy additional cars for his branch before the airport opens a new runway that will increase air traffic and passenger numbers. Information about the runway opening will very likely not occur as a message within the dynamic process solution supporting the branch manager's work.

Consequently, without going into details of the internal behavior of a component, specifying the component communication already defines important aspects of the component behavior.

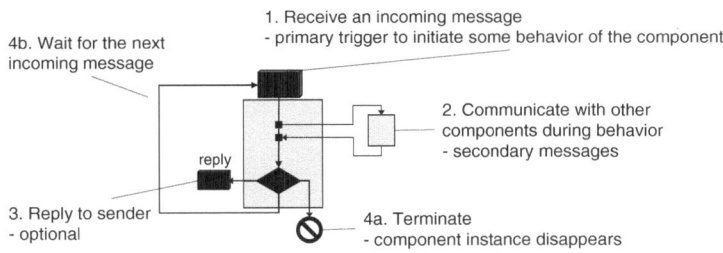

Fig. 4. Processing cycle of a communicating component

- **Primary Trigger:** What types of incoming messages do the component trigger?
- **Reply**: What reply will a component provide?
- **Secondary Incoming messages:** What types of additional incoming messages can the component handle during a processing cycle?
- **Secondary Outgoing messages:** What types of additional outgoing messages can the component send to obtain further information to compute its reaction to the primary trigger?
- **Visibility scope:** Which other components exchange messages with the component?
- **Published information:** What information does the component publish once a processing cycle is completed?
- **Internal memory**: Does the component maintain an internal memory? In the case of the object life cycle, the internal memory would capture the object state; however in the case of an actor, the memory can be much more comprehensive, storing message and activity histories for example.

This information can be specified using a BPMN collaboration diagram. In a concrete design, one detailed BPMN model would be drawn for each business rule, for each activity of an actor, and for each state transition of an object life cycle. Keeping a separate component for each activity and state transition leads to simpler collaboration diagrams. Furthermore, it will facilitate the configuration and modification of dynamic BPM solutions built with the proposed approach.

The next subsections specify the information from the preceding list for each component type using an abstract form of a collaboration diagram with disconnected model elements within a pool, which does not define details of the component behavior.

4.1   Business Objects

A business object consists of the data object stored in the data base and the associated life cycle, which is deployed as a communicating component instance in the runtime. In the ePVM implementation, life cycle instance and data base object are linked by using the component instance id as the primary key of the data object in the data base. Remember that the method proposes that the object life cycle controls all manipulation of a business object in the corresponding data base to ensure consistency between the life cycle instance and the object.

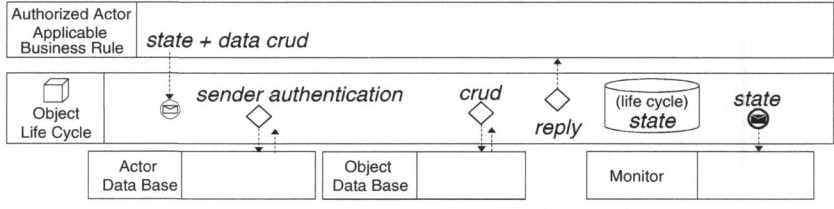

Fig. 5. Object life cycle component interaction

Figure 5 shows an abstract BPMN collaboration diagram for object life cycles. The primary trigger is represented as the incoming message of the start event, wheras any published information is captured as the outgoing message of the end event of the pool representing the component. The reply of a component to the primary trigger is an outgoing message to the component that is the source of the trigger. The secondary messages occur between the start and end event. They originate from a decision gateway if they are optional. The data store symbol represents the internal memory of the component and shows the information that the component stores internally. The visibility scope is represented by other components as the originators or receivers of messages. They are represented by additional empty pools.

The primary trigger of a life cycle component is a message from an authorized actor or an applicable business rule. A life cycle component is not aware of other components. It can return a reply to the sender of the trigger message, but not actively contact other components. The life cycle component keeps the current state in its internal memory. While transiting to the next state, it can send a message to the actor data base to request the authentication of the sender of the primary trigger. Sender authentication turned out to be an important capability of object life cycles in a dynamic BPM solution. A life cycle receives messages from other components about which it does not know any details nor is it even aware that these components exist. As these messages can arrive in arbitrary order from arbitrary senders, a life cycle should have the capability to verify that a sender is indeed authorized to send a state/data change. For this purpose, the life cycle is aware of data sources (or services), where it can query information about the sender. Accepted changes are persisted in the corresponding business object in the data base. Upon finishing the processing cycle, the object life cycle will publish the state to a monitor. The monitor can record the state change history of the object and optionally inform other components that are interested in a state change of the object.

Figure 6 shows a concrete example of a BPMN collaboration diagram for the state transition from *acquired* to *available* of the car object life cycle shown in Figure 2.

The car life cycle expects certain state update messages in certain internal states, e.g., only the new state *available* is expected in state *acquired*. These messages can only originate from the car park assistant authorized to do the initial inspection on a newly acquired car. Consequently, when the *available* message arrives, the car queries the actor data base to confirm that the responsible car park assistant has sent this message. Only upon authentication and if the data passes other internal validations, will it transit

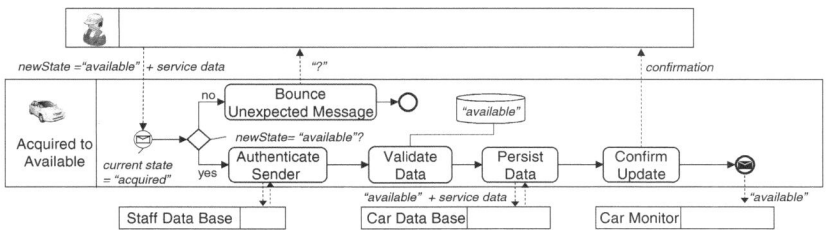

Fig. 6. Collaboration model for the state transition from *acquired* to *available* of the car object life cycle

to the new state and persist the state and data change in the car data base. If the update was successfully confirmed by the data base, the car life cycle sends a confirmation to the sender of the primary message and publishes the new state to the car monitor. If the component receives another message than the one expected, it interprets it as an error. Two reactions were implemented in the case study: swallow the message and do nothing or bounce it back to the sender as a reply. The latter is shown in the example model. Both give the sender a chance to realize that its communication might not have achieved the desired behavior at the receiving side: In the first case, no confirmation or notification of a state change will eventually happen. In the second case, the object directly notifies the sender through the reply.

4.2   Business Rules

A business rule component is triggered by a timer event, a query from another business rule, a request from an actor, or a state of an object life cycle, see Figure 7.

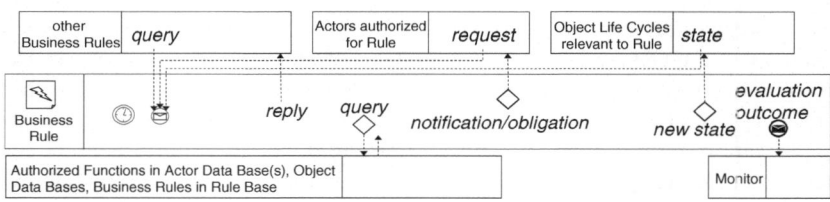

Fig. 7. Business rule component interaction

To evaluate its outcome, the rule can query information about business objects from the corresponding data bases for which it is authorized and use authorized functions from the actor data base(s) as well as other business rules. It responds with a reply to a query or request, can send a notification or obligation to an actor, or can send a state update to a business object life cycle component. Upon completion of its processing cycle, it publishes its evaluation outcome to allow the monitor to maintain a history of executed business rules.

Figure 8 shows the model for the rule *"at the end of each day, cars are assigned to reservations for the following day"*. In this model, the concept of "end of day" has

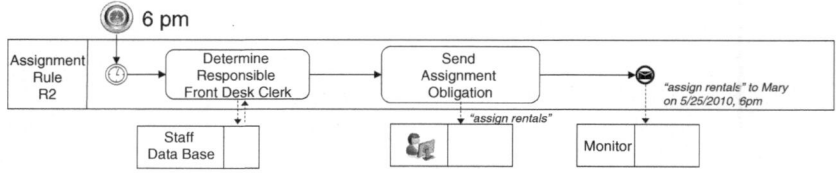

Fig. 8. Collaboration model for the rental assignment rule

been mapped to the specific timer event of 6 pm. When the timer fires, the rule first determines the actor for whom it has to establish the obligation. Then it sends a specific obligation message to this actor.

4.3   Actors

Actor components have much stronger communication capabilities than object life cycles and rules. They can spontaneously initiate some of their activities, while others might be triggered by an incoming message. Figure 9 illustrates both options, showing two types of start events.

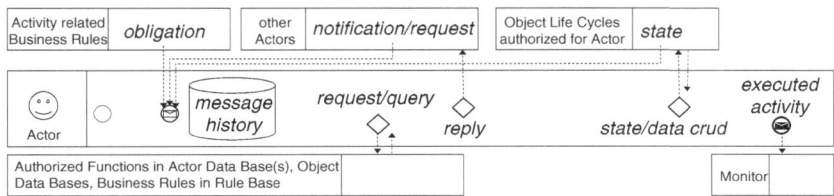

Fig. 9. Actor component interaction

A triggering message can be an obligation established by a business rule, a notification or request from another actor, or a state information from a life cycle component. An actor keeps a message history in its internal memory, which allows the actor to correlate information to perform a complex behavioral activity. During the activity, actors can manipulate business objects by interacting with their life cycles or consult business rules. An actor can also send a request to other actors, which will trigger processing cycles for these actor components. An activity can result in the intention to change the state and data of a business object, which results in the corresponding message to the object life cycle component. The life cycle component may reply with a confirmation, which the actor receives as a secondary incoming message. An actor will usually return a reply to any request received. Upon finishing the processing cycle, the actor publishes information about the executed activity to the monitor, which allows the monitor to maintain the activity history of the actor. Business processes resulting from the message-coordinated activities of several actors can be mined based on such activity histories.

Activities usually work with object life cycles and business rules. For example, the branch manager initiates a new car life cycle when buying a car. A front desk clerk queries the actor data base to find out which car park assistants are currently on duty. To find a suitable upgrade for a customer, she consults business rules that represent the upgrade policies of the EURENT car rental company. Figure 10 illustrates how business rules, object life cycles, and activities play together in an end-to-end business process. The process starts with the branch manager buying a new car. A new car life cycle in the initial state *acquired* is instantiated through the *buy car* activity. The publication of this initial state state by the car triggers a technical inspection by the car park assistant

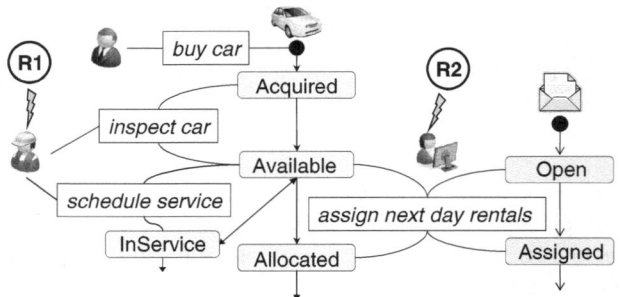

Fig. 10. Actors, business rules, and object life cycles in an end-to-end business process

taking the car to state *available*. Whenever a car enters this state, business rule (R1) *"each car must be serviced every three months or 10,000 kilometers, whichever occurs first"* is evaluated, which may trigger a service. The car park assistant will execute the *schedule service* activity to respond to this obligation, but decides about the time frame himself, optionally monitored by another business rule not shown in Figure 10.

A car can also be rented to a customer when being in state *available*. The rental process begins when the front desk clerk chooses to assign this car to a rental contract by executing the *assign next day rentals* activity, which is initiated by rule (R2) *"at the end of each day, cars are assigned to reservations for the following day"*.

The *assign next day rentals* activity is of particular interest as it causes a synchronized state transition of two business objects, namely, the rental contract transiting from state *open* to state *assigned* and the rental car transiting from state *available* to state *allocated*. BPMN collaboration models for this activity are shown in Figures 11 and 12.

The *assign next day rentals* activity is triggered by the *assignment rule R2*. The concrete timing for the triggering message was specified by the rule component model in Figure 8. The activity consists of two main tasks. First, the front desk clerk must obtain the next day rentals from the rentals data base. Note the use of a collection data object *rentals*, which comprises all *rental* objects of state *open* for which the pick-up date equals the date of the next day. As this task is only a look-up not changing data,

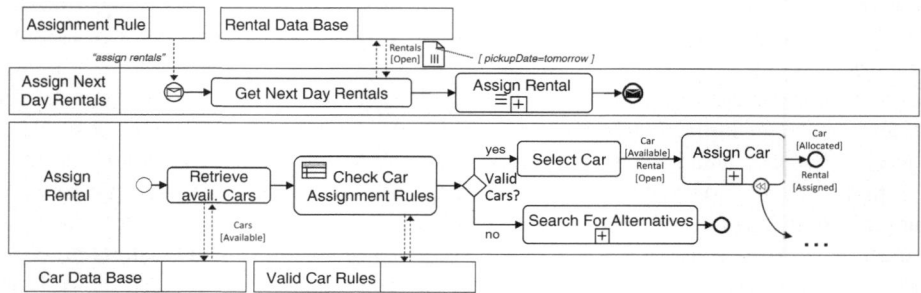

Fig. 11. Collaboration model of the *assign next day rentals* activity

redirection through the object life cycle is omitted from the model. Second, a car is assigned to each open rental contract in the *assign rental* multi-instance subprocess. Details of this subprocess are further spelled out in the lower part of Figure 11.

The *assign rental* subprocess begins by retrieving available cars from the car data base for each *rental* object in the *rentals* collection. As the assignment should happen the evening before the rental car is picked up, cars on rent or cars in service are unlikely to be good candidates, i.e., only cars in state *available* are retrieved. Next, a conventional business rule task *check car assignment rules* is executed to query rules to determine whether an available car is also a valid choice for the rental reservation, e.g., is from the car group requested by the customer. If this is the case, one of the valid cars is selected and finally assigned to the rental contract. Otherwise, an alternative car assignment, e.g., an upgrade, has to be found by the *search for alternatives* subprocess, of which details have been omitted. The *assign car* subprocess encapsulates the synchronized state transition of the car and rental business objects.

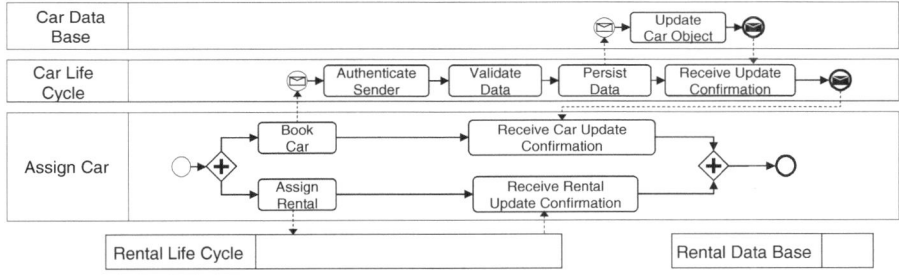

Fig. 12. Collaboration for the synchronized state transition of the *car* and *rental* business objects

The collaboration model in Figure 12 specifies the message communication between the *assign car* subprocess, the car and rental life cycles, and the car and rental object data bases. It shows details of the car data base and car life cycle at the top, whereas the rental data base and rental life cycle are only shown collapsed at the bottom of the figure. To achieve the correct coordinated behavior of the car and rental life cycles, the *assign car* subprocess sends update messages to the car and rental life cycle components. The life cycle components in turn persist the updates in the corresponding car and rental data bases. Confirmations are sent from the data bases to the corresponding life cycles and from there to the *assign car* subprocess. If one of the confirmations does not arrive after a certain period of time, the front desk clerk (and probably also the car and rental life cycles) should initiate a compensating/error-handling activity, which is not further detailed in the example collaboration models.

5 Next Steps

The precise separation of activities, life cycles, and business rules helps users of this methodology to focus on the exact behavior of each element. However, they will benefit from additional support to better understand the complexity of the resulting composite behaviors. This section identifies selected research problems that have to be tackled to take the proposed methodology from the current prototype status to a mature and industrial-strength solution approach.

Extensions of BPMN 2.0: The collaboration diagrams in Figures 5, 7, and 9 have been used in this paper to illustrate the communication behavior of each component type. In a specific dynamic BPM solution, business analysts will draw concrete collaboration diagrams like those in Figures 6, 8, and 11. In an initial modeling phase, such diagrams could be drawn as private processes showing only the incoming and outgoing messages. For this purpose, it would be desirable to visualize conditional message flow.

To make the human actors explicit who own the various activity components, lanes have been used. It needs to be investigated in greater depth whether the limited support for role modeling in BPMN 2.0 is sufficient for the proposed methodology. In addition, the relation to UML state machines must be further investigated. For example, Figure 6 shows a textual annotation that refers to the current state of the car life cycle, which influences how the incoming message is handled. The integration of the state attribute of a state machine model like the one in Figure 2 with its representation in the collaboration diagram representing a specific transition is currently an open question.

Refinement of the modeling methodology: When applying the integrated approach to the EURENT domain, the more than 100 business rules were grouped by object life cycle states. Structural business rules govern specific life cycle states, whereas operative business rules initiate, prevent, or govern state transitions. This helped to quickly bring structure into the large rule set and to detect inconsistencies. It would be desirable to evolve this aspect of the methodology into a systematic support for the structuring and consistent description of business rules, which helps business analysts to systematically transform textual business rules into operational communicating components.

Further investigations are required to confirm whether the methodology covers all aspects of case-handling solutions. It clearly provides information based on object states with business rules governing the activity flow. Within these boundaries, actors can self-initiate activities. Decoupling of authorization and distribution can also be achieved. The authorizations of a component are defined by other components with which it can interact. The work distribution in a business process can be organized by controlling the communication flow, possibly with the help of monitors.

An open question is the measurement of the quality of a dynamic process solution, for example, how the "right" granularity of components can be found. Further study also needs to be devoted to more advanced object interaction patterns, such as many-to-many synchronization relations between business objects, where, for example, one object can spawn an unknown number of "subordinate" object instances at runtime whose life cycles need to evolve to certain states before the spawning object can transit. In the spirit of the method proposed here, it would not be desirable that the object life cycles achieve such a coordination by communicating directly with each other. Instead, it seems that business rules would be the natural way to capture such complex forms of object coordination.

Tooling capabilities: In a concrete design, one detailed BPMN model would be drawn for each business rule, for each activity of an actor, and for each state transition of an object life cycle. Maintaining the consistency and linkage between the various BPMN models, e.g., organizing the set of collaboration diagrams representing all possible activities of one actor or linking a collaboration diagram to each transition of a state machine model representing the life cycle of a business object would be a main task for a BPMN editor supporting this modeling methodology. It also seems possible that BPMN *conversation* diagrams could be used to visualize component interactions and to generate different views, such as, for example, all components that can receive or send a particular message or all activities that affect a particular object state.

Static analysis and simulation: Simulation and static analysis capabilities become very important as it can be expected that solutions are incrementally build starting from a small set of components or are configured by selection from a set of pre-built components. In this context, scenario-driven agile development and testing as proposed in the context of embedded systems design [6] seems to be a very promising approach for the design of dynamic BPM solutions. A play engine that allows business analysts to initiate and observe the behaviors (scenarios) that result from a set of BPMN collaboration components and that automatically constructs the overall end-to-end process would facilitate the dynamic BPM solution creation. In addition, tools should possess analytical capabilities. These may range from detecting that there is no component in a solution that receives a message sent by another component to becoming aware of races between activities, for example, the service scheduling and the rental of the car.

Mapping to existing or novel runtime engines: Conventional BPM engines that support component communication or alternative approaches such as [7] can be used for implementation. Some commercial BPM suites already support business rules as communicating components based on the Service Component Architecture (SCA) standard.

Actor interfaces: The design of user-friendly interfaces for humans working in such a dynamic BPM solution is another interesting area of work. Users see the following types of activities: activities that are always possible, e.g., create rental, activities that are enabled or required by specific object life cycle states, e.g., contact the customer of a delayed car, activities that result from rule obligations, e.g., assign cars, and activities that reflect requests from human actors, e.g., reassignment of a rental contract.

6   Related Work

So far, business rules and business processes have been integrated in a static way whereby a process invokes a rule (or rule set) at a specific point in the flow. Gartner [8] recently described an entire spectrum of integration scenarios for business processes and rules, but also emphasizes how little is understood about these scenarios. The approach in [9] to embed rule-based control and compliance objectives in a business process design shows how complicated a tight integration can be, leading to a complicated process design and rather motivates the clear separation between rules and processes in the methodology discussed in this paper. The two most relevant standards, SBVR [3] and BPMN [2], emphasize the need for integration, but also declare it as out of scope. BPMN only provides the so-called business rules task, which allows a BPMN engine to synchronously invoke a set of business rules over the data of the process and use the result of the rule evaluation to decide on the further branching of the process. This solution of using rules as automated decisions is widely accepted, but as the authors of [10] emphasize, it is important to make the reasons explicit upon which obligations arise and cease, and to allow actors to decide which actions to take as a consequence of the obligation instead of directly invoking a specific action based on a computed rule outcome.

The increasing need for a more explicit treatment of business information in business processes has led to a widespread agreement that business objects should play a more prominent role. For example, ARIS emphasized early on that an event represents the fact that an information object has taken on a business-relevant state which is controlling or influencing the further procedure of the business process [11]. However, the ARIS method has not paid explicit attention to the evolution of the business object states. The artifact approach [12] carried the idea on to a sophisticated approach of state machines that capture the life cycle of business objects. Instead of defining business processes explicitly, the approach assumes that a process results from the interaction of the object life cycles with each other. Recently, a proposal about how to combine classical processes with business object life cycles was published [13], but the life cycles lose much of their autonomy and operational power, because a very closely-coupled, transaction-oriented manipulation of life cycles by processes is envisioned.

Explicit dependencies between life cycles are investigated in [14], where they are used to synthesize coordination structures between the transitions of different life cycles. The consistency of object life cycles and business process models as well as methods to derive one from the other are discussed in [15].

The integration of business rules and object life cycles has only received little attention so far. The authors of [16] follow an event-condition-action paradigm and use rules to synchronously control the transition to the next state. The work in [17] studies a variant of the artifact approach where services, which correspond to activities in the methodology presented in this paper, are annotated with expressive pre- and postconditions and sets of rules are synthesized that dynamically generate a workflow from service invocations to achieve a given goal formulated over attributes of an artifact. The work is very interesting; however, the high computational complexity of the synthesis problem already under the limitation that no artifact interactions are permitted and services can only be invoked once by a workflow, leaves many open questions regarding the practical applicability and extensibility of this work.

An initial treatment of business processes and business objects as communicating components can also be found in [18,19]. Constraint-based approaches [20,21] have recently been proposed to capture dependencies between data, activities, and business rules using temporal constraint languages. Although promising, the proposals show a complete paradigm shift in how BPM solutions are modeled and pose new problems, such as the need for an efficient and complete constraint solver for the proposed languages.

In contrast to approaches that assume activity/service annotations with complex pre- and postconditions, the methodology as described here keeps the "precondition" of an activity limited to a single message and encodes the knowledge that guides activity execution in separate business rules. This makes the methodology more naturally aligned with business rules modeling as it is common today and avoids complicated declarative specifications of activities, which has been identified as a major knowledge engineering bottleneck by the Artificial Intelligence (AI) community [22]. The AI community proposes a rather orthogonal approach based on negotiation and intelligent decision making between autonomous agents as the foundation of dynamic BPM [23,24,25] with business processes remaining essentially implicit. At this stage, it seems that negotiation patterns as well as process and activity patterns as, for example, described in [26,27] can be adopted by the proposed methodology in the form of best practices and pre-defined component templates.

7 Conclusion

A modeling methodology for dynamic BPM solutions is introduced that treats *business rules*, *actors*, and *business object life cycles* as equal partners in a loosely coupled system that interact through message exchanges, but have different capabilities. The clear distinction of the three component types makes it possible to separate component behavior from component interaction and facilitates the implementation of business patterns. BPMN collaboration diagrams capture the communication and behavior of each component. This methodology puts much more emphasis on messages and events in collaborations than conventional business process modeling does. The importance of scenario-driven development and testing is briefly discussed and identified as one of the open research challenges, not only for BPMN tool builders.

Acknowledgment. I would like to thank Thomas Weigold for his support in using ePVM, Mark Linehan for his help in getting started with SBVR, the Zurich BIT team and Ksenia Wahler, ipt, for interesting discussions and valuable feedback, and the anonymous reviewers for their encouraging comments.

References

1. van der Aalst, W.M.P., Weske, M., Grünbauer, D.: Case handling: a new paradigm for business process support. Data and Knowledge Engineering 53(2), 129–162 (2005)
2. Object Management Group: Business Process Model and Notation (BPMN), Version 2.0, OMG document number dtc/2010-05-03 (2010)
3. Object Management Group: Semantics of Business Vocabulary and Business Rules (SBVR), Version 1.0, OMG document number formal/2008-01-02 (2008)
4. Business Rules Group: Defining business rules - what are they really? Final Report (2000)
5. Weigold, T., Kramp, T., Buhler, P.: Flexible persistence support for state machine-based workflow engines. In: 4th Int. Conf. on Software Engineering Advances, pp. 313–319. IEEE, Los Alamitos (2009)
6. Harel, D., Marelly, R.: Specifying and executing behavioral requirements: the play-in/play-out approach. Software and System Modeling 2(2), 82–107 (2003)
7. Abiteboul, S., Bourhis, P., Marinoiu, B.: Efficient maintenance techniques for views over active documents. In: 12th Int. Conf. on Extending Database Technology, pp. 1076–1087. ACM, New York (2009)
8. Sinur, J.: The art and science of rules vs. process flows. Research Report G00166408, Gartner (2009)

9. Sadiq, S., Governatori, G., Namiri, K.: Modeling control objectives for business process compliance. In: Alonso, G., Dadam, P., Rosemann, M. (eds.) BPM 2007. LNCS, vol. 4714, pp. 149–164. Springer, Heidelberg (2007)

10. Abrahams, A., Eyers, D., Bacon, J.: An asynchronous rule-based approach for business process automation using obligations. In: ACM SIGPLAN Workshop on Rule-Based Programming, pp. 93–103 (2002)

11. Davis, R.: Business Process Modeling with ARIS. Springer, Heidelberg (2003)

12. Nigam, A., Caswell, N.: Business artifacts: An approach to operational specification. IBM Systems Journal 42(3), 428–445 (2003)

13. Nandi, P., et al.: Data4BPM: Introducing business entities and the business entity definition language (BEDL), IBM developerWorks (2010)

14. Müller, D., Reichert, M., Herbst, J.: Data-driven modeling and coordination of large process structures. In: Meersman, R., Tari, Z. (eds.) OTM 2007, Part I. LNCS, vol. 4803, pp. 131–149. Springer, Heidelberg (2007)

15. Küster, J., Ryndina, K., Gall, H.: Generation of business process models for object life cycle compliance. In: Alonso, G., Dadam, P., Rosemann, M. (eds.) BPM 2007. LNCS, vol. 4714, pp. 165–181. Springer, Heidelberg (2007)

16. Linehan, M.: Ontologies and rules in business models. In: 11th Int. Conf. on Enterprise Distributed Object Computing (EDOC), pp. 149–156. IEEE, Los Alamitos (2007)

17. Fritz, C., Hull, R., Su, J.: Automatic construction of simple artifact-based business processes. In: 12th Int. Conference on Database Theory (ICDT). ACM Int. Conf. Proc. Series, vol. 361, pp. 225–238. ACM, New York (2009)

18. van der Aalst, W., Barthelmess, P., Ellis, C., Wainer, J.: Workflow modeling using Proclets. In: Scheuermann, P., Etzion, O. (eds.) CoopIS 2000. LNCS, vol. 1901, pp. 198–209. Springer, Heidelberg (2000)

19. Redding, G., Dumas, M., ter Hofstede, A., Iordachescu, A.: Modelling flexible processes with business objects. In: IEEE Conference on Commerce and Enterprise Computing (CEC), pp. 41–48. IEEE, Los Alamitos (2009)

20. van der Aalst, W.M.P., Pesic, M.: DecSerFlow: Towards a truly declarative service flow language. In: The Role of Business Processes in Service Oriented Architectures. Dagstuhl Seminar Proceedings, vol. 6291 (2006)

21. Wu, Q., Pu, C., Sahai, A., Barga, R.: Categorization and optimization of synchronization dependencies in business processes. In: 23rd Int. Conf. on Data Engineering (ICDE), pp. 306–315. IEEE, Los Alamitos (2007)

22. Barták, R., McCluskey, L.: Introduction to the special issue on knowledge engineering tools and techniques for automated planning and scheduling systems. Knowledge Enginering Review 22(2), 115–116 (2007)

23. Taveter, K., Wagner, G.: Agent-oriented enterprise modeling based on business rules. In: Kunii, H.S., Jajodia, S., Sølvberg, A. (eds.) ER 2001. LNCS, vol. 2224, pp. 527–540. Springer, Heidelberg (2001)

24. Jennings, N., Norman, T., Faratin, P., O'Brien, P., Odgers, B.: Autonomous agents for business process management. Applied Artificial Intelligence 14(2), 145–189 (2000)

25. Weyns, D.: A pattern language for multi-agent systems. In: Joint Working IEEE/IFIP Conference on Software Architecture 2009 and European Conference on Software Architecture (WICSA/ECSA), pp. 191–200. IEEE, Los Alamitos (2009)

26. Hruby, P.: Model-Driven Design Using Business Patterns. Springer, Heidelberg (2006)

27. Thom, L., Reichert, M., Iochpe, C.: Activity patterns in process-aware information systems: basic concepts and empirical evidence. Int. Journal of Business Process Integration and Management 4(2), 93–110 (2009)

A Concept for Spreadsheet-Based Process Modeling

Stefan Krumnow[1] and Gero Decker[2]

[1] Hasso-Plattner Institute at the University of Potsdam,
Prof.-Dr.-Helmert-Str. 2-3, 14482 Potsdam, Germany
stefan.krumnow@student.hpi.uni-potsdam.de
[2] Signavio GmbH,
Goethestr. 2-3, 10623 Berlin, Germany
gero.decker@signavio.com

Abstract. When it comes to the understanding and usage of business process models, there exists a gap between process modeling method experts and domain experts. This paper proposes an approach to close this gap by using simple spreadsheet-based representations of process models. For this, different approaches for the modeling of business processes in spreadsheets are introduced and evaluated. Transformations from and into BPMN 2.0 are investigated and on basis of these results, a round-tripping concept for process models is developed. Finally, a prototype is introduced that proves the applicability of this concept.

Keywords: Process modeling, Spreadsheets, Excel, Transformation, BPMN, Round-Tripping, Version Management.

1 Introduction

Graphical process model notations such as the BPMN provide established and successful ways for the specification or documentation of intended or existing processes. They offer accepted syntaxes and at least semi-semantics. Thus, a graphical model supports the communication between those people who understand its notation (method experts) and can be used in analysis- or implementation-projects.

However, these notations tend to be rather unintuitive for people who do not know them. Even worse, software tools that can be used for the creation and maintenance of such graphical models are complex and require special user skills in most cases. Therefore, the majority of people in a company will not be able or allowed to participate in the work with process models. Hence, thorough interviews of a persons who completely know and understand a certain process are required as it could not influence the process model otherwise.

In order to close the gap between BPM method experts and domain experts, other representations of a process model can be helpful. A process could be modeled in a simpler notation using a tool that is easy to use. One approach to realize this idea is the usage of spreadsheets. Spreadsheets offer a simple and (especially

J. Mendling, M. Weidlich, and M. Weske (Eds.): BPMN 2010, LNBIP 67, pp. 63–77, 2010.
© Springer-Verlag Berlin Heidelberg 2010

in business) well-known way to organize data in rows and columns. Moreover, they can be edited by wide-spread tools like Microsoft Excel or OpenOffice Calc, which is even free of charge. So a spreadsheet-based process model could be easy to understand and handle for most people who use computers for work.

This paper investigates how a process can be represented synchronously in a spreadsheet-based model as well as in a graphical model. For this purpose, first it has to be identified how processes can be modeled in spreadsheets in a way that is as simple but also as expressive as possible. Then, transformations between graphical notation and spreadsheets have to be defined. In order to really close the gap between process and domain experts, the transformation algorithms will then be used in a synchronization mechanism that allows for a model to roundtrip between both worlds. Hereby, the main idea is to export graphical models into easier spreadsheets that can be modified by domain experts and be imported back into the process modeling tool. All results developed in this paper are applied in a prototype that finally shows how the round-tripping can work in praxis.

2 Existing Tools and Related Work

There exist already some process-modeling tools that use spreadsheet-based modeling approaches.

First of all, Symbio Process[1] should be mentioned. This tool, developed by a small German company, uses Microsoft Excel enhanced with VBA macros in order to model processes completely in a spreadsheet. The modeler can add activities or events by adding rows to the spreadsheet. Columns are used to represent properties of the elements represented by rows. The first two properties of each element are its name and type. Then, there are properties for assignments, organizations, input, outputs, involved systems and form templates. Moreover, Symbio Process allows the modeler to integrate control flow structures into the spreadsheet. Each element has a predecessor property. Besides activities and events there are also connectors and process interfaces available in order to support links between different models and branching and joining behavior within one model. Symbio Process offers a transformation into graphical process models (e.g. into BPMN and EPCs) that are exported as images.

Another interesting tool is ARIS Express[2], a freeware version of ARIS. ARIS Express is a graphical modeling tool for processes and organizations that offers (amongst others) the notation of EPCs. Within the modeling canvas, the modeler can open a spreadsheet editor called SmartDesign. This editor offers, similar to Symbio Process, a view on the current process fragment, whereby activities and events are represented by rows and their properties by columns. Changes on process data within the SmartDesign editor (e.g. the renaming of activity or the adding of a new input document) are directly taken over to the graphical process model.

Unlike Symbio Process, ARIS Express does not support any control flow structures in its SmartDesign editor: Only sequences of activities and events can be

---

displayed and edited. In cases of EPCs that include branching and joining this causes the SmartDesign editor to only show a process fragment around a selected element.

Although it does not perfectly fit into the field of spreadsheet-based solutions, also Lombardi Blueprint[3] should be mentioned here. Lombardi Blueprint offers web-based process modeling in different modeling perspectives such as a graphical canvas or a more simple grouping perspective. Here, activities are grouped into milestones and carry properties. The user interface of Lombardi Blueprint does not use spreadsheets but drag and drop containers and dialogs. However, the application offers a very simple Excel export, which lists all activities in separate rows and names their associated milestones.

Although there exist several commercial process-modeling tools that use spreadsheet-based modeling approaches, the topic is mostly ignored in the BPM research community: [6] and [7] show that spreadsheets are often used in process analysis, controlling and simulation. In [2], the authors even propose a transformation of process models into spreadsheets for the purpose of simulation. However, business process modeling taxonomies such as [4] do not cover spreadsheets as a way of modeling processes.

Model transformations and the synchronization of models are more general topics, e.g. addressed in the field of model-driven engineering [5].

3   Catalog of Approaches

As the last section has shown, the existing tools share one common characteristic: Rows in a spreadsheet are used to represent elements of interest (such as activities or events) while columns are used to represent property types. But there are many differences between the tools as well. These differences can be categorized by regarding the whole problem of what to express in a spreadsheet in different dimensions:

First of all, there is the dimension regarding the question of which **elements** to support. In ARIS Express, activities, events and process interfaces (links or subprocesses) can be represented by a row. In Symbio Process, also branching and joining connectors are supported as elements of interest. Compared with graphical notations, both approaches share a limitation: Elements such as data objects or organizations can only be specified as property values but not as own elements with own properties. Therefore, e.g. an input document of an activity can only be described by a string and not further defined, e.g. by an xml schema.

In a second dimension, the question of how **properties** are represented is investigated. In Symbio Process, there are a number of predefined properties that can (for each element) store one string value. In ARIS Express, it is also possible to have multiple assignments for one property (e.g. two input documents for one activity). Again, there is a shared limitation as only a predefined set of properties exists in both solutions. In an enhancement, the definition of customized properties in the spreadsheet itself could be possible.

---

[3] https://blueprint.lombardi.com/index.html

The third dimension covers the question of which **relations** between elements can exist. ARIS Express only supports a very simple relation of elements: the sequence defined by the order of rows. Symbio Process on the other hand allows for the modeling of control flow relations between certain elements. If a tool allowed the modeling of data or organizational elements as own rows, it would also have to offer a possibility to model association e.g. between activities and data objects.

In the following subsection, we will develop a catalog of differently expressive approaches to model processes using spreadsheets based on categorization presented above. We will start with the most easy but also less expressive approach and will then increase difficulty and expressiveness.

3.1   The "Simple Sequence" Approach

As the existing tools presented in section 2 still aim for business process method experts as users, this first approach tries to be as simple as possible: Only the modeling of sequences of activities is supported. Activities have a number of property types that can be assigned once or several times per activity. These multiple assignments of properties are necessary, e.g. for activities that have two input data objects.

Figure 1 shows how this approach can be modeled using an UML class diagram. The Activity class inherits its behavior from a class Row that encapsulates all the characteristics an Excel row has, including its row number and the ordering within a spreadsheet. Besides a name and a description, activities can define input and output data as well as performers and their organizations.

A difficulty of this approach exist in the realization of multiple properties in a spreadsheet. If an customized spreadsheet editor is used (as in ARIS Express), the cells can be enabled to contain lists of values. If Excel is used, the problem is much harder. Here, comma-separated strings could be used. Alternatively, a line-break-separated string could be used, which leads to a list layout for a multiple property.

3.2   The "Branching" Approach

As most process models contain not only sequences but complex control flow structures, the second approach adds some new elements and, more importantly a successor property to each element. In order to support branching and joining

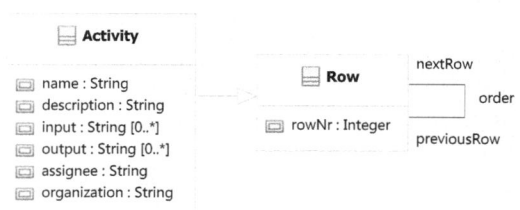

Fig. 1. The "Simple Sequence" approach modeled as UML class diagram

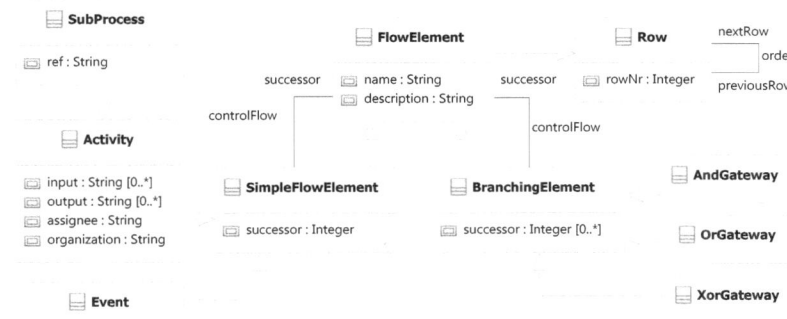

Fig. 2. Elements of the "Branching" approach modeled as UML class diagram

within one model, gateways and events are neccessary. As in BPMN, the basic three gateway types (Parallel, Exclusive and Inclusive) have different execution semantics. Events can be used to express reached states in a process model. More important in our case, they can be used to model different conditions after branching gateways. Finally, subprocesses are activities, that have an own process model describing their internal behaviour.

Figure 2 shows the extended UML class diagram that models the structural characteristics of the second approach. The abstract class FlowElement now holds name and description of an element, while BranchingElement and SimpleFlowElement areintroduced to cover the successor property. For branching elements, more than one successor can exist. Several successors can, e.g. , be realized by using a comma-separated string of row numbers as property value. If no successor is specified, implicitly the next row's element is interpreted as successor. If that row is empty, the current element has to be an end-element.

One problem that occurs now is the identification of the element type a certain row represents. This could be corrected by introducing a new property to each row that holds the type of the represented activity. As this requires the user to understand the differences between the types, a smoother approach can be taken: Depending on the given name, the position within the process' flow and used properties, the type of an element can be deducted (by humans as well as

	A	B	C	D	E	F	G	H	I
1	Name	Description	Input	Output	Assignee	Organization	IT System	Successor	...
2	Receive Order								
3	Pack ordered goods		Order	Package Case Doc	Storage D	MyShop	ERP		
4	XOR	Is amount > 25€?						5,6	
5	Add vouchers to package		Order Package Case Doc	Archived State	Shipment D	MyShop		6	
6	Ship Package		Package Case Doc	Package Case Doc	Shipment D	MyShop	DHL Support		
7									

Fig. 3. Cutout of an example spreadsheet-based process model

computers). Here, only an example in figure 3 should illustrate this. In sections 4 and 5, the idea is explained in more detail.

The example shows a short sequence of activities that are needed to handle an order, e.g. in an online shop. First, the goods have to be packed. Then, if the volume of order is high enough, some vouchers will be added to the package and finally the package is shipped. In order to make "Add vouchers" optional, a splitting exclusive gateway that can be identified by the beginning of its name ("XOR") is placed in the model.

3.3   The "More Properties" Approach

Although the second approach already supports a wide range of features a process model might have, there is still one gap to close: The modeler has no support to specify properties, e.g. of an input document or an assigned role. As mentioned before, this is due to the fact that all those elements are just represented by simple strings. In order to fix this problem, elements like data objects or organizations could be represented by own rows that are linked to activities similar to the control flow that links two control flow elements.

Moreover, the modeler is limited in the number and types of properties a element might have. This can be corrected by giving him the oppurtunity to define own customized property types by adding new columns to his spreadsheet. In figure 4, the extended UML class diagram that also covers the third approach is shown.

But there is also a downside of this approach. Although we gain more expressiveness when it comes to the behavior and structure of attached elements, the spreadsheets are getting more complex: There are a lot of associating links connecting an activity row to attached element rows, even if for the attached element only the name is modeled.

Fig. 4. Elements of the "More Properties" approach modeled as UML class diagram

This section showed three approaches that allow the easy modeling of processes in spreadsheets. In order to rank the approaches, two aspects have to be regarded: Firstly, the approach should be as easy to use as possible. The ranking of this aspect alone is simple, as the difficulties to use a certain approach are increasing from I to III. Secondly, the approach should be applicable for as many processes as possible. The next section investigates how spreadsheets of the different approaches can be transformed from and into existing graphical notations. This leads to an empirical research that examines what percentage of an huge model repository is covered by each approach, which might help to build up a ranking of approaches.

4 Transformations

The question of transformations between graphical models and spreadsheet models has several motivations: As already explained in the last section, it helps to rank different approaches by their expressiveness. But far more important is the fact that spreadsheet-based process models cannot exist in an isolated environment: Graphical notations such as the BPMN have proven to be successful for those that work with process models in their everyday's work. So, in order to benefit from the established techniques for graphical models as well as from the easiness of spreadsheets, both worlds have to be transformable into each other.

Therefore, this section will show how spreadsheet-based model can be transformed from and into BPMN. We will start by introducing the general transformation algorithm (that could also be used for other notations like EPCs) before we show which elements of the BPMN are transformable.

4.1 General Algorithms

In order to transform a graphical model into a spreadsheet, we need a transformation T. The following code snippet shows how T can be implemented:

```
nodeArray = getNodesInSequentialOrder(model) for each node in
nodeArray:
  if isSupportedElement(node):
    createRow(node)
for each node in nodeArray:
  if isSupportedElement(node):
    for each connectedNode of node:
      if isSupportedRelation(node, connectedNode):
        addRelation(node, connectedNode)
```

For a natural ordering of elements within the produced spreadsheet, firstly the nodes have to be sequentialized. Thereby, pairs of two nodes that are connected by control flow, whereby the source has only one outgoing flow and the target has only one incoming flow are placed next to each other in an list. By iterating over this list, the spreadsheet can be created. Here, four functions have be implemented. The first two functions check whether an element or an elements' relation is valid. The third creates rows, while the fouth adds connections between rows (or data values to rows).

There are two loops used in order to show that for the creation of a link between a node and its connected node, this connected node needs to be represented by an existing row. The second loop could, of course, be merged into the first one in order to gain performance, but this would lead to a more complex code snippet. However, as the sequentializer as well as the two loops have a linear time consumption, the whole transformation lies in $O(n)$. Please also note that connected nodes might also be nodes that have a vertical (containment-) relation with the currently regarded node.

In order to transform a valid spreadsheet into a graphical notation, a transformation T^{-1} is needed. Here, a similar algorithm can be used:

```
for each row of spreadsheet:
  if isSupportedRow(node):
    createElement(row)
for each row of spreadsheet:
  if isSupportedRow(node):
    for each property of row:
      // this includes the implicite order of rows
      if isLink(property) && isSupported(node, property):
        addRelation(row, property)
```

Of course, this transformation requires a layouting algorithm for the generated graph. As this is a complex problem itself, this paper assumes the existence of such an algorithm that can be applied for the resulting model. Like T, also T^{-1} can be performed in linear time.

4.2 BPMN Transformation

After looking at the general transformation algorithms, we can now inspect which parts of the BPMN 2.0 [1] can be transformed into spreadsheets and vice-versa. Table 1 shows which elements and relations can be transformed into the different spreadsheet approaches and backwards using uni- or bi-directional arcs .

As the table shows, all elements of the three approaches can be transformed into BPMN elements. Thereby, every row of a spreadsheet is interpreted as an activity except if one of the following criterias is met:

- In II and III: If the name starts with AND, XOR or OR, the element is a gateway.
- In II and III: If the element has no predecessor/successor, it is a start event/end event.
- In II and III: If the ref property is set, the element is a collapsed subprocess.
- In III: If the element is connected through a typed relation (e.g. input), the element's type can be deducted from the relation.

Table 1. Transformability of BPMN elements and relation

	Approach I	Approach II	Approach III
Task	↵	↵	↵
All Events, Collapsed Subprocess	⊘	↵	↵
Parallel-, Databased & Eventbased Exclusive-, Inclusive-Gateway	⊘	↵	↵
Collapsed Subprocess	⊘	↵	↵
Pool, Lane	⊘	↵[5] + ⬆	↵[5] + ⬆
Data Object	⊘	↵ + ⬆[6]	↵
Subprocess, Collapsed pool, Complex Gateway, Group, Annotation	⊘	⊘	⊘
Sequence flow	↵ + ⬆[4]	↵	↵
Message flow	⊘	⊘	⊘
Associations (directed and undirected) with Data	⊘	↵ + ⬆[7]	↵ + ⬆[7]
Containment in Pool and Lane	⊘	↵	↵[8] + ⬆
Attachement of boundary events	⊘	⊘	⊘

In the other direction, there exist a lot of BPMN elements that cannot be transformed into either one of the approaches, e.g. subprocesses and collapsed (black-box) pools but also the descriptive elements groups and annotations.

As table 1 also shows, message flows cannot be transformed into spreadsheets at all. Since message flows are used to model the interaction of different processes, they are a pretty complicated construct and can therefore not be pressed into a spreadsheet-based model that aims for simplicity. Moreover, attached boundary events cannot be transformed and there exist restrictions for containment in pools and lanes as well as for associations.

4.3   Statistical Analysis

The identification of BPMN elements and relations that can be transformed into spreadsheets gives us the opportunity to investigate how expressive the three approaches really are. This can be done by investigating how many BPMN models of the huge Oryx repository [3] can be transformed into the different approaches from section 3.

---

[4] If maximal one outgoing and maximal one incoming flow occurs per element.
[5] Attribute Organization (=pool) has to be specified when Assignee (=lane) is used.
[6] If data object is connected to a task and no properties are modeled.
[7] Undirected and bidirectional relations will appear equal after transformation.
[8] If maximal one Organization (=pool) and maximal one Assignee (=lane) is defined.

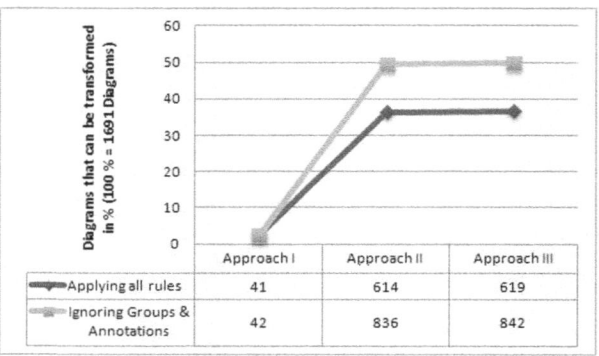

Fig. 5. Results of the empirical analysis on transformanility of 1691 BPMN models

The Oryx repository contains roughly 2500 BPMN process models. Before these can be used in an analysis, small test models and models with a big number of syntactical errors had to be filtered as they could falsify the results. On the basis of a filtered set of 1691 meaningful BPMN models, we conducted an analysis that investigates how many of the models can be transformed into either of the three presented approaches.

For this purpose, every meaningful model has to run through a check algorithm that identifies all the approaches this model can be transformed into without any loss of information according to table 1. Similar to the algorithm shown in section 4.1, the check algorithm looks for each element and relation which approaches can cover it. Thereby, when regarding those element's properties, some difficulties occur due to the fact that Oryx stores element properties not only if they are set by the user but also if there exist default values for them. Thus, it is not possible to detect all properties that the users really intended to set, which might slightly falsify the results presented in figure 5 regarding the usage of unsopported properties.

As the diagram shows, only 36% of the models can be transformed without loss of information. This could of course be corrected by inventing approaches that are even more expressive than approach III. But on the other side, these approaches would then loose the advantage of being simple enough for domain experts and not only for process modeling experts. Therefore, from this analyzation, we can conclude that one-to-one mappings between graphical and spreadsheet-based modeling notations will not help closing the gap between those two groups. Rather, we need transformations that accept the loss of certain information. Figure 5 shows that only by ignoring groups and annotations almost 50% of the models become transformable. Interestingly, approaches II and III are almost equally expressiveness for this set of test data.

The next section will now show how graphical models and spreadsheet models can be synchronized even if the graphical model contains complex constructs that cannot be represented in an easy-to use spreadsheet model.

5   Round-Tripping

In order to close the gap between domain and method experts, a model should be able to roundtrip between a graphical and a spreadsheet representation. There are two use cases for such a roundtrip:

A certain process is modeled in a graphical notation (e.g. BPMN) using a modeling tool (e.g. Oryx). The method expert likes to let a domain expert having a view on the model and maybe correcting some faults. Therefore, he exports the process into a simple spreadsheet, sends it to the domain expert who can simply edit it with Excel. Afterwards, the domain expert sends the spreadsheet back to the method expert who can import the spreadsheet back into his system.

In the second use case, the process in initially modeled in a spreadsheet and then send to a method expert who imports it to his system. Afterwards, the same roundtrip as in the first use case might of course take place. Section 4 already showed how spreadsheets can be transformed into graphical models, which is sufficient for the second use case.

But for the first use case, we cannot use these straight-forward transformation algorithms. This is due to the fact that graphical models are more expressive than easy-to-understand spreadsheets are: If we call the transformation from graphical into spreadsheet models T and the vice-versa transformation T^{-1}, we observe the following problem: By applying the round-tripping transformation $T^{-1}(T(g))$ on a graphical model g, the outcome of the transformation might differ from its input, as T accepts the loss of information.

Moreover, the method expert would be forced to not edit the graphical model until he gets the edited spreadsheet back. Otherwise he will end up having to manually merge different versions into one model. The next section will show how both problems of the round-tripping use case can be solved by applying techniques from version control systems.

5.1   Concept

In figure 6, a concrete example for a model roundtrip is shown. In the upper right section of the image-matrix, a revision graph shows, which stages the model runs through. The other sections show relevant cut-out of the different model's versions and representations. The roundtrip starts (as described before) with a graphical model that contains the task "Pack ordered goods" with a catching boundary event (I). This event leads to an exception handling in case that the ordered goods are not available anymore.

As we have already seen in section 4.2, these events cannot be transformed into spreadsheets. Nevertheless, the model can still be transformed into a spreadsheet that still contains the task and the event (II) by using the before-presented transformation. Only information of an concrete connection between the two elements is lost here. Therefore, the domain expert receives a model perspective abstracting from certain constructs but still containing information on all used nodes as well as their properties, which is suitable for understanding and performing minor changes.

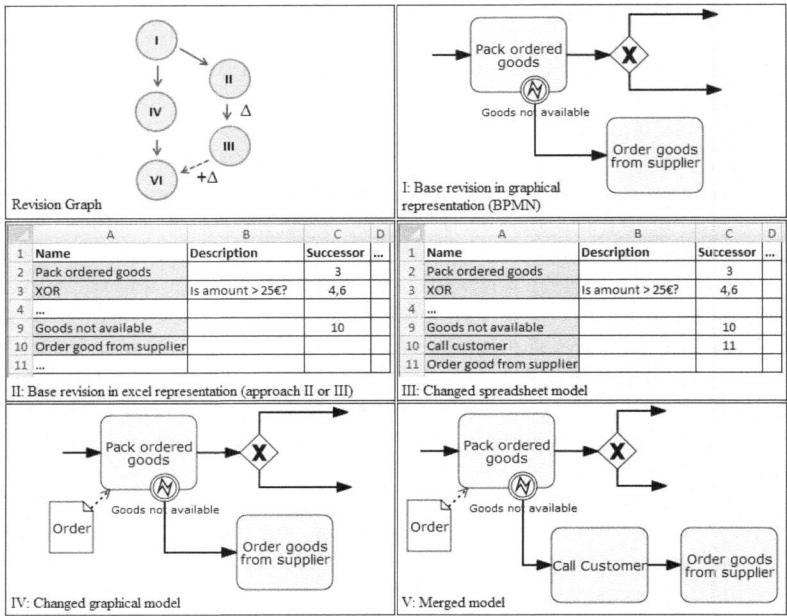

Fig. 6. Examplary roundtrip of a order process model

Now, the domain expert discovers that in cases where not all ordered goods are available, the customer is also called in order to inform him. So he adds a new task to the spreadsheet (III). Meanwhile, the method expert starts adding additional data to the graphical model (IV). When the domain expert sends the spreadsheet back, the method expert should be able to simply import the changes performed on the spreadsheet into his graphical model (V).

Although the spreadsheet does not specify how "Pack ordered goods" and the "Goods are not available" event relate to each other, they are still connected after the round-tripping transformation and the subsequent flow has even been extended by the domain expert who only saw his incomplete spreadsheet.

This is possible by not using T^{-1} to acquire the new version g^* as before. Instead, we use a merging transformation M: We get $g^* = M(g^{head}, \Delta)$, where Δ denotes the set of performed changes on the spreadsheet and g^{head} is the newest graphical revision. Δ can be acquired by comparing the base spreadsheet revision e^{base} (which is the same as $T(g^{base})$) and the newest spreadsheet revision e^{head}. This brings us to:

$$g^* = M(g^{head}, comp(T(g^{base}), e^{head}))$$

Thus, a graphical process modeling tool can merge back changes performed on a spreadsheet only with the input of the (graphical) model's head and base revision as well as the current version of the spreadsheet. Of course, this tool must also offer the transformation T in order to transform graphical models into spreadsheets. Moreover, for the import of spreadsheets into the tool, T^{-1} should be supported. In the next section, we will present a prototypical implementation of such a tool that is build on basis of the modeling tool Oryx [3].

5.2 Protype

In order to demonstrate the proposed round-tripping features, we implemented a prototype within the graphical process modeling tool Oryx. It meets all the requirements that derive from the use cases presented above.

There are several design decisions that have to be made when implementing such a round-tripping system, starting with the selection of supported graphical notations as well as the spreadsheet approaches: Among EPCs (whose transformation has not been shown here), the prototype was supposed to support BPMN. When it comes to supported approaches, the main focus should be to deliver easy-to-understand spreadsheets. Due to that, only approaches I and II from section 3 were supposed to be supported by the prototype, especially as the third approach did not offer much more expressiveness in the analysis.

Thus, the developed prototype supports the round-tripping of BPMN and EPC models with "Simple Sequence" and more advanced "Branching" spreadsheets. To do so, it has to offer the transformation T for both notations into two spreadsheet variants. The implementation of this export transformation is according to the transformation that has been presented in section 4 but allows information to get lost. Therefore, also diagrams with branching and joining control flows can be exported into the "Simple Sequence".

Moreover, the prototype has to support the reimport of both spreadsheet types into existing models. Section shows 5.1, that this requires a merging algorithm M that merges a change set into an existing model as well as a comparator function *comp* that detects the performed changes on a spreadsheet. To do this, the comparator needs to have the spreadsheet's revision at the time of its export.

There are different approaches possible to provide this version: The spreadsheet base revision could be provided by the user when reimporting the newest revision. Also, this revision can be stored within the system. Then, only the base revision number has to be specified. A third option is to only store a graphical representation of the base revision and apply transformation T a second time, when the reimport is performed. As this approach saves storages and as there already exist Oryx backends that store all former revisions of a graphical model, the prototype uses this last approach. It thereby stores the needed information to identify the base revision (model identifier and revision number) directly in the exported spreadsheet. Thus, the user does not have to provide additional information besides the changed spreadsheet when reimporting it.

The comparator *comp* now takes two spreadsheets as input and detects all changes. Changes are the adding of a new row, the deletion of an existing row or the modification of a existing row. A row can be modified by adding a new property value or by removing or changing of an existing value. Thereby, also successorship is handled as a property value.

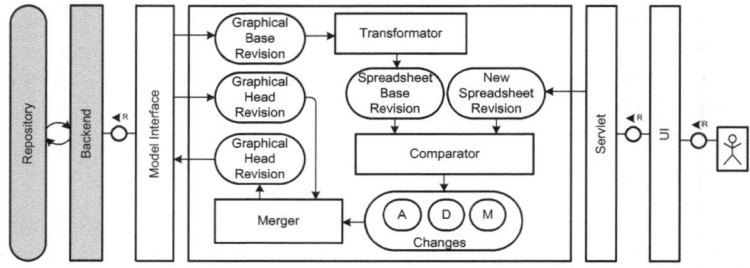

Fig. 7. Prototype's components and internal dataflow when reimporting

Now, the merger M integrates the detected changes into the graphical model. The deletion of a row results in the deletion of the corresponding BPMN element with all its attached objects such as data. If the element was not only removed from the spreadsheet but also from the graphical model in the meantime, it can of course not be deleted. Such a conflicting change can only be logged and displayed to the user after the reimport.

The adding of a new element raises the problem of where to place it. The prototype takes a simple layouting approach here: If the element's predecessor lays in the same lane as the new element is supposed to be put in, the element is placed on the right next to it and all later elements in this lane are shifted to the right. If the predecessor lays in another lane, the element is placed into his lane using same x-coordinate its predecessor has. All subsequent activities within the lane are moved to the right.

For modified elements, properties are added, changed or removed. For the adding of a property value that has to be represented graphically (e.g. an input data object), again a simple layouting algorithm is used that finds a free spot next to the element. Figure 7 gives a final overview over the prototype's components and its dataflow when reimporting a spreadsheet.

6   Conclusion

This paper proposes an approach to close the gap of understanding between process domain experts and process model method experts by using spreadsheets. The idea is to support the round-tripping of a model in both representations in order to benefit from graphical notations' expressiveness as well as from spreadsheets' simplicity. The paper investigates all relevant aspects of the idea's realization and finally presents a prototype demonstrating the approach.

First of all, it classifies different approaches how to represent business processes in spreadsheets. Therefore, existing solution are investigated as well as further ideas developed. Then, the transformability of the different approaches from and into BPMN 2.0 is shown. The results and a large model repository are used to conduct an empirical analysis on the expressiveness of the different approaches. On top of these transformations, a concept for the round-tripping of models is developed and realized.

This concept allows to also exchange process models, which cannot be completely transformed, with domain experts using simple spreadsheets. Using techniques similar to version control approaches, changes within the spreadsheet can be fully automatically merged into the original BPMN model, as long as no conflicting changes (i.e. changes on the attribute or conflicting removals) have been applied there. In order to show the applicability of this concept, we developed a prototype that allows for the round-tripping of BPMN (but also EPC) process models using spreadsheets.

In future work, we would like to investigate to which degree the abstraction from certain modeling constructs during spreadsheet creation is acceptable for domain and method experts, which might result in restrictions regarding the set of supported process models. Moreover, the prototype is going to be enhanced by using a customized spreadsheet editor or macros with the spreadsheet. Also, the usage of a better auto-layouter is desirable, as only simple layouting techniques have been used so far.

References

1. Business process model and notation (bpmn), version 2.0 - beta 1. Technical report, Object Management Group (OMG) (August 2009)
2. Bradley, P., Browne, J., Jackson, S., Jagdev, H.S.: Business process re-engineering (bpr)—a study of the software tools currently available. Comput. Ind. 25(3) (1995)
3. Decker, G., Overdick, H., Weske, M.: Oryx — an open modeling platform for the bpm community. In: Dumas, M., Reichert, M., Shan, M.-C. (eds.) BPM 2008. LNCS, vol. 5240, pp. 382–385. Springer, Heidelberg (2008)
4. Giaglis, G.M.: A taxonomy of business process modeling and information systems modeling techniques. International Journal of Flexible Manufacturing Systems 13 (2001)
5. Giese, H., Wagner, R.: Incremental model synchronization with triple graph grammars. In: Nierstrasz, O., Whittle, J., Harel, D., Reggio, G. (eds.) MoDELS 2006. LNCS, vol. 4199, pp. 543–557. Springer, Heidelberg (2006)
6. Kettinger, W.J., Teng, J.T.C.: Business process change: a study of methodologies, techniques and tools. MIS Q. 21(1) (1997)
7. Tumay, K.: Business process simulation. In: WSC 1996: Proceedings of the 28th Conference on Winter Simulation. IEEE Computer Society, Los Alamitos (1996)

Managing Complex Event Processes with Business Process Modeling Notation

Steffen Kunz, Tobias Fickinger, Johannes Prescher, and Klaus Spengler

Humboldt-Universität zu Berlin
Institute of Information Systems
Spandauer Str. 1, 10178 Berlin, Germany
{steffen.kunz,tobias.fickinger,johannes.prescher,
klaus.spengler}@wiwi.hu-berlin.de

Abstract. Complex Event Processing – the identification of event patterns in event streams, the analysis of their impact, and the execution of corresponding actions – is gaining momentum within various research disciplines and business areas. However, one of the major problems associated with Complex Event Processing is its lack of usability, especially the complexity of its management, preventing its wide diffusion and adoption by users. This usability issue is addressed in this paper by applying Business Process Modeling Notation as graphical support for the definition of complex event patterns.

Keywords: Complex Event Processing, Event-Driven Business Process Management, Event Processing Language, Business Process Modeling Notation.

1 Introduction

Complex Event Processing (CEP) is a technology for obtaining relevant situational information from distributed services in real-time (or almost real-time) by selection, aggregation, and event abstraction for generating higher level complex events of interest. [1] [2].

An event in the context of CEP is defined by [2] as an object that represents, encodes, or records an (real-life) event, generally for the purpose of computer processing. A simple-level event stream can, e.g., originate from sensor networks or RFID readers. When detecting specified complex events, further activities, like an Event Processing Workflow (EPW), can be triggered [3].

The definition of event patterns for the detection of complex events is done with the Event Processing Language (EPL) [4, p. 33]. This language is similar to the Structured Query Language (SQL) for database queries, but has additional elements to define special statements required for CEP. The main difference between both languages is that "streams replace tables as the source of data with events replacing rows as the basic unit of data" [4, p. 33]. Today, there exists no standard for an event processing (query) language, though different solutions are available (e.g., [5], Esper [4] and Oracle [6] provide their own EPLs, Microsoft uses LINQ [7]).

J. Mendling, M. Weidlich, and M. Weske (Eds.): BPMN 2010, LNBIP 67, pp. 78–90, 2010.

The problem CEP is currently facing is its lack of usability preventing its wide diffusion in real life. This paper tries to increase the usability by providing a graphical, user friendly presentation of EPL statements through Business Process Modeling Notation (BPMN) artifacts, thus supporting a wider application and dissemination of CEP technology. In our approach, we especially focus on the Esper EPL standard [4]; other EPL standards can easily be integrated, too.

In the following Section, we present a reference scenario, which will be used to exemplify and motivate our approach. In Section 3, we provide an EPL-2-BPMN mapping approach which allows representing EPL statements through BPMN 2.0 [8] artifacts. Section 4 addresses related work and Section 5 discusses our current and future work. Finally, in Section 6 we give a conclusion.

2   Reference Scenario

In the following paragraphs, a reference process, which has relevant characteristics for modeling complex-event detection through CEP, and corresponding EPL statements are described.

2.1   Logistics of Perishables

Perishables are goods with high requirements towards the transportation process. In order to ensure that consumers obtain high quality goods, an integrated cold chain is required.

In our scenario, we assume that the goods are tagged with a low priced RFID chip storing the Electronic Product Code (EPC) [9] thus providing a unique ID for each good. Various sensors distributed along the supply chain provide data about the environment and the condition of the goods. Depending on the respective processes, these data streams are either stored, retrieved, and processed or directly processed in real-time by the CEP engine to identify complex events in simple event streams.

We use the following high level reference process to discuss how BPMN models can be used for displaying events that should automatically be detected and handled within the supply chain. In our approach, we put the focus on one specific object of the product class fish filet (see Figure 1): The path through the supply chain starts with the production of the fish filet. Then, it is packed in a box which is stored in the manufacturer's cold store. After an order has been received from the retailer, the box is shipped to the retailer by a logistics service provider. At the retailer site, the object is first stored in a cold store, then moved to a cold shelf on the shop-floor. Due to brevity, we do not display intra-logistics such as transportation to/from the retailer or manufacturer distribution hub

The requirements towards transportation throughout the whole supply-chain are as follows:

- The temperature should be below -18 ° C.
- The humidity should be below 80% (to avoid condensation within the packages).

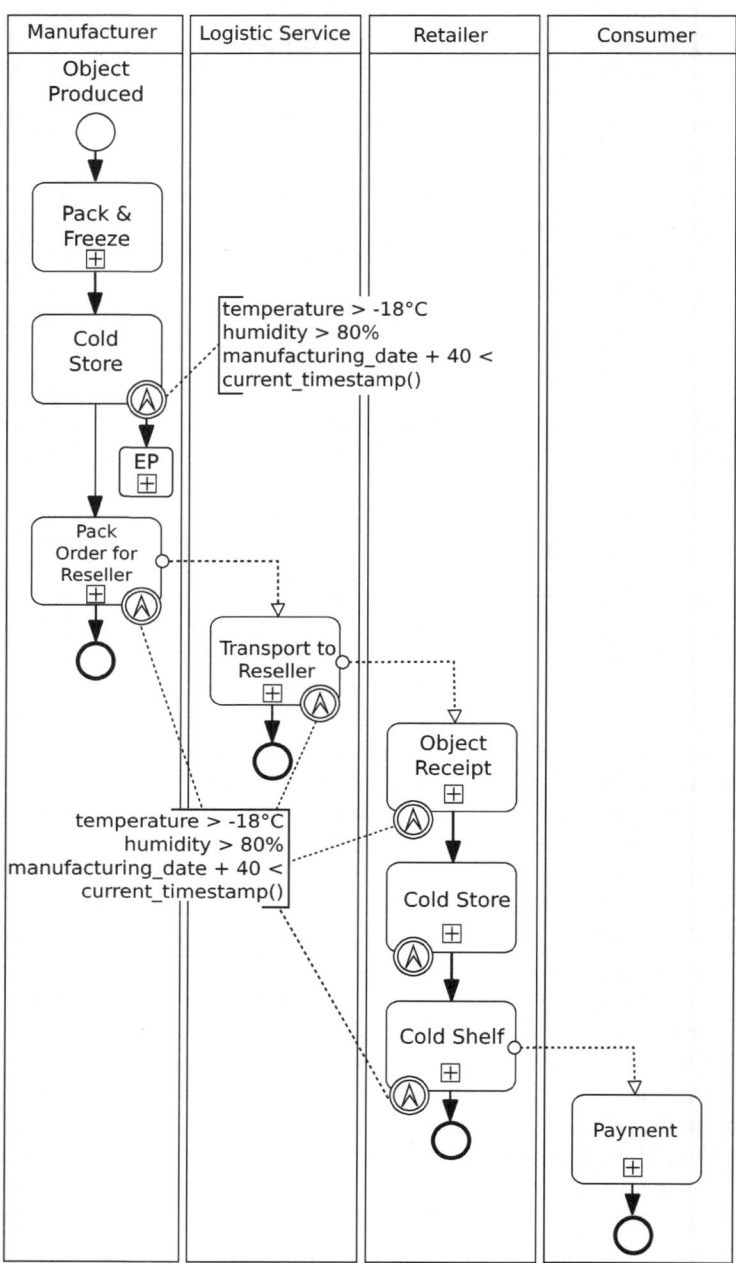

Fig. 1. Reference Process: Logistics of Perishables

- The product should not be older than 40 days (manufacturing_date + 40) during transportation.

In Figure 1, the reference process is divided into pools depicting the different domains. At certain waypoints of the supply chain, requirements towards the transportation of the object are represented by escalation events [8, pp.243] which trigger an Escalation Process (EP) if these requirements are not fullfilled. Due to a lack of space, we only presented one EP in the reference process, which is attached to the task cold store in the manufacturer's pool. In this case, the escalation could be to replace the spoiled fish filet.

2.2   Reference Process EPL Statements

The above mentioned requirements towards can be described in EPL code:

```
SELECT id,location,temperature,humidity, current_timestamp()
FROM frozenGoodStream
WHERE temperature > -18
OR humidity > 0.8
OR manufacturing_date + 40 < current_timestamp();
```

The code can be divided into three clauses: The first clause (SELECT) defines the specific attribute data – in this case, the ID of the specific object, the physical location in the logistics process with the associated temperature and humidity, and the current time. The data source (FROM), here named "frozenGoodStream", could be a real time event stream from sensor networks or a historical event streams from distributed event repositories. The conditions that merge different simple events into a complex event can be found in the third (WHERE) clause.

3   The EPL-2-BPMN Approach

The goal of the following approach is to map the complete EPL syntax elements to existing BPMN 2.0 artifacts. Major challenges involved in the reference process are those that arise from exceptions due to their influence on the process flow.

Since the EPL syntax can be divided into three major parts (SELECT, FROM, and WHERE), we suggest a mapping focusing on these elements (see Figure 2). A detailed description of the mapping is given in the next sections. EPL syntax elements such as aggregate, grouping, or join functions will not be mapped by BPMN artifacts, but by their corresponding attributes.

3.1   Scan Activity Represented as BPMN Service Task

In order to model an event-driven business process with BPMN, the modeler needs to place an event-sensitive BPMN element at a certain location in the model to trigger the scanning of the event stream for an event pattern.

According to the BPMN 2.0 [8] specification, we use the *Service Task* element as a derivation from the *Task* element [8, pp.135]. The Service Task element in

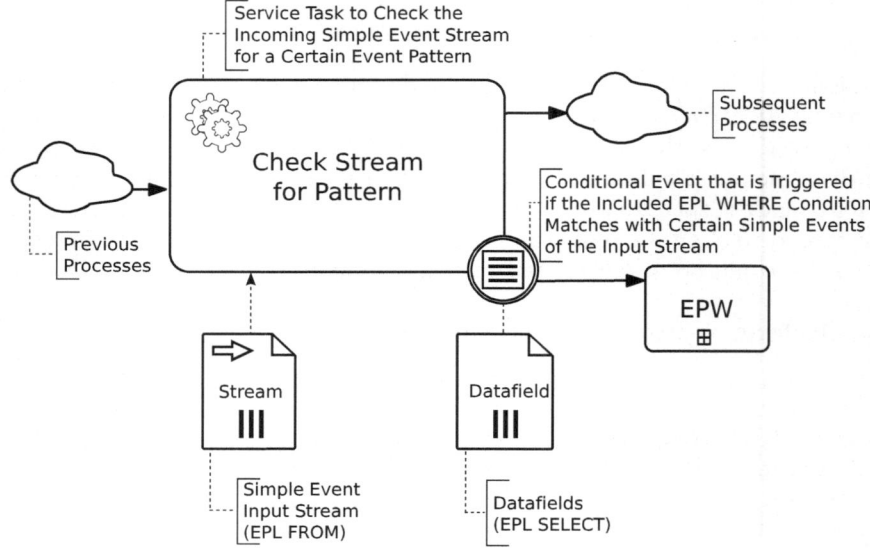

Fig. 2. EPL-2-BPMN Approach

Fig. 3. BPMN Service Task Element

BPMN represents a task that uses some sort of service, e.g., an automated application. We intend to use the Service Task artifact for the scan-stream activity (Check Stream for Pattern) in our model (see Figure 3). Besides standard attributes that a Service Task inherits, such as ID (derived from BaseElement) or name (derived from FlowElement), we also use specialized attributes [8, p.137] as described in Table 1.

Table 1. Attributes of Service Task

Attribute Name	Usage
Implementation	This attribute specifies the technology (e.g., RFID Scanner) that will be used to receive the incoming stream.
operationRef	This attribute specifies the operation (e.g., scan task) that is invoked by ServiceTask.

Fig. 4. FROM Statement presented as BPMN Data Input Collection Element

3.2   FROM Statement Represented as BPMN Data Input Collection

Since the Service Task represents the main activity of scanning the simple event stream, we propose placing the EPL elements (SELECT, FROM, and WHERE) in the context of this task. For our further considerations, we define the incoming simple event stream as information provided by the sensor environment or from an event repository. Since the *Data Object* represents the primary construct for modeling an information flow within a BPMN process [8, pp.183], we use this structure for the mapping of the incoming stream. Based on the huge amount of simple events provided by the stream, an extension of the Data Object as *Collection* [8, p.185] is necessary.

The Data Object Collection is refined by the BPMN *InputOutputSpecification* class [8, pp.190] emphasizing that the scan task requires a stream input to run. This class provides the *Data Input* specification, which specifies that a particular kind of data will be used as input. Finally, the Data Input Collection is connected by an outgoing *Data Association* with the Service Task. The Data Input Collection (see Figure 4) typifies the FROM EPL statement.

For connecting the JOIN functionality of EPL to at least two streams, we apply the BPMN construct of *Input Sets* [8, pp.196]. Here all Data Input Collections associated with a single Service Task relate to a single Input Set via the corresponding reference attribute. The characteristic of the concatenating Input Set then represents the intended JOIN requirement.

In addition to the standard attributes, a Data Input Collection inherits, e.g., the ID derived from BaseElement, we define specialized attributes as described in Table 2 [8, p.193]. Here, we neglect the derivation from ItemAwareElement and focus on the relevant attributes.

3.3   WHERE Condition Represented as Boundary Conditional Event

Depending on the result of the stream scan, there are two possible sequence flows out of the Service Task mentioned above: (a) The normal flow in the boundaries of the environmental business process, which will be activated if the event condition has not been fulfilled. (b) An exceptional flow to provide an applicable reaction to a specific event (pattern). The exceptional flow is represented by the EPW [3]. Based on these two options, the Service Task of our framework is extended by a boundary event, which allows two outgoing associations [8, p.232]. Regarding the normal flow, the association will be connected directly from the Service Task to the next element of the standard process. In case of a detected complex event, the association leads from the boundary event to the EPW.

Table 2. Attributes of the Data Input Collection

Attribute Name	Usage
name	Name of the DataInput (e.g., frozenGoodStream). This name will be used as part of the FROM statement and should consist of the technical name of the incoming event stream.
inputSetRefs	At least one InputSet is defined for each DataInput (in terms of JOIN requirement several DataInputs need to refer to a single InputSet).
inputSetWithOptional	The ServiceTask as the associated Activity requires the InputSets. Therefore no optional InputSet should be declared.
inputSetWithWhileExecuting	To determine that ServiceTask can evaluate its DataInput while executing, all InputSets should be declared here.
isCollection	Stated as true to represent the multiplicity of incoming simple events.

Table 3. Attributes of Conditional Event

Attribute Name	Usage
condition	The WHERE conditions are placed as expressions in a formal language.
cancelActivity	Based on the requirements the normal sequence flow can be set as interrupted (true) if the ConditionalEvent is being detected.

Fig. 5. WHERE Condition Represented as Intermediate Interrupting and Non-interrupting Conditional Event

We chose the *Conditional Event* [8, pp.241] as an intermediate event attached to an activity boundary [8, p.232]. The Conditional Event type inherits a conditional expression that can be tested as being either true or false. The event triggers an exception flow if the condition becomes true. Furthermore, the conditional expression will typify the WHERE-Conditions of the EPL statement. The particular conditions can be modeled by using the expression attribute of the ConditionalEventDefinition ([8, pp.241], see Table 3).

We suggest using both *interrupting* and *non-interrupting* Conditional Events (see Figure 5) depending on the business process requirements [8, p.253]. The interrupting Conditional Event has the function to interrupt or even terminate a complete process, if this is necessary as reaction to a complex event. On the contrary,

Table 4. Attributes of DataObject Collection

Attribute Name	Usage
isCollection: Boolean	Stated as true to represent the multiplicity of required data fields.

the non-interrupting Conditional Event could comply with the defined business process, but trigger another task, e.g., a data collection for later data analysis.

3.4 SELECT Statement Represented as Data Object Collection

In the context of complex event processing with EPL, certain information is gathered by scanning the incoming simple event stream for event patterns. The EPL statement includes the SELECT construct to define which data fields will be selected if a complex event is detected. This information (e.g., EPC or other ID, position, etc.) is forwarded to the EPW for further processing. The corresponding BPMN construct for transferring information to linked subprocesses is described by *Data Object* [8, pp.183]. Since several SELECT fields can be part of an EPL statement, we suggest to extend the Data Object by using Collections (see Figure 6) [8, p.185] – similar to the FROM statement. However, contrary to the Data Input of the FROM statement, the Data Object for SELECT fields is neither declared as input of the framework, nor directionally associated to a framework element. Rather, we propose to use an undirected association connected with the conditional boundary event. This construct enables the triggered Conditional Event to transmit selected data fields to the subsequent EPW. Furthermore, the undirected association keeps the original intent of SELECT fields within the framework as a data selection compared to an incoming information stream [8, p.56].

Fig. 6. SELECT Statement represented as Data Object Collection

3.5 EPL-2-BPMN Summary

In our approach, we mapped the EPL elements SELECT, FROM, and WHERE to the corresponding BPMN artifacts (see Figure 2). The suggested approach, including all relevant BPMN classes, attributes, and associations, can be described by an UML class diagram (see Figure 7), which may be used as the basis for the implementation of a EPL-2-BPMN modeling tool. All artifacts in the context of our approach follow the original BPMN 2.0 specification [8]. We propose

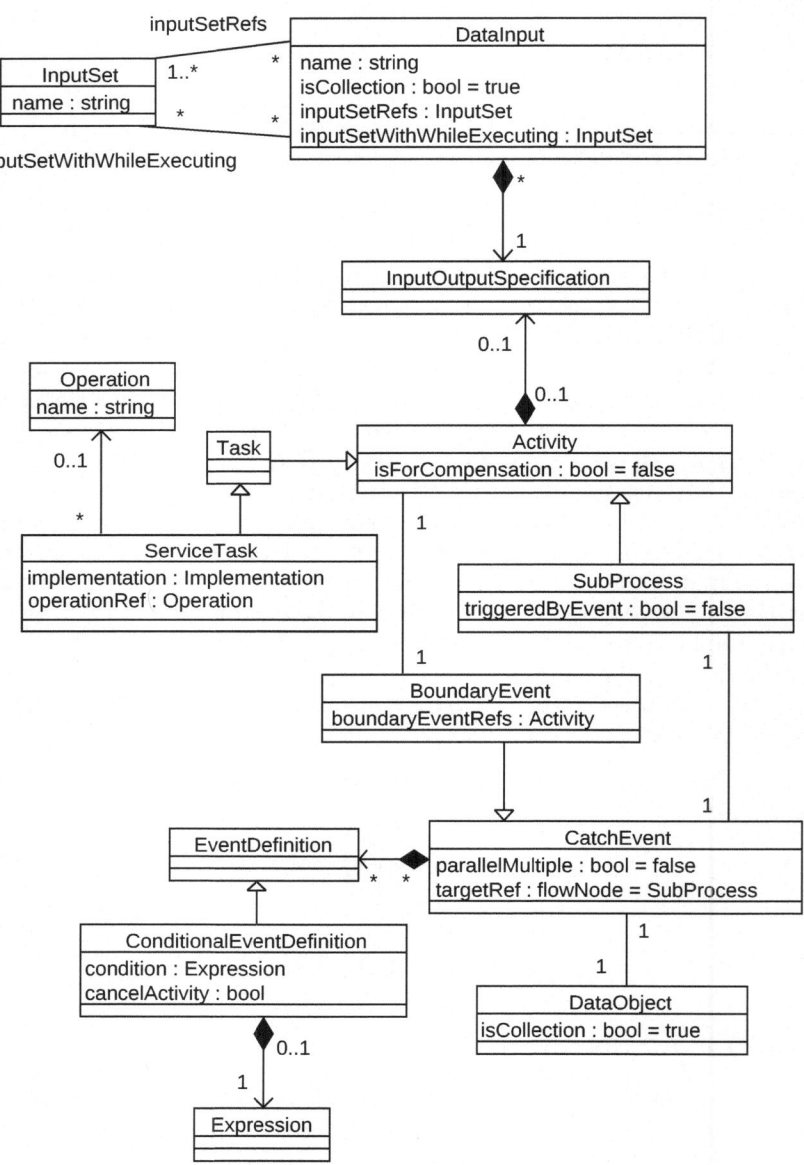

Fig. 7. Class Diagram of the EPL-2-BPMN Approach

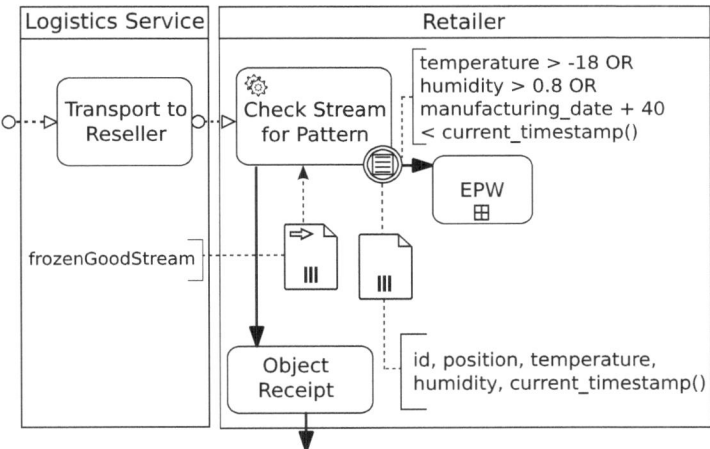

Fig. 8. EPL-2-BPMN Approach Integrated into the Reference Process

to use default values and data types of certain attributes as well as the precise application of the environmental class definitions and associations as described in the UML class diagram.

3.6   Reference Process Integration

One of the most reasonable locations to control the quality and quantity of goods in a supply chain are the transitions from one business domain (here a BPMN pool) to another. In the given reference process, these edges are arranged between manufacturer, logistics service provider, retailer, and customer.

To exemplify the integration of the proposed approach into the process model, we chose the transition from logistics service provider to retailer, where the goods are transported to the retailer. During transportation, storage, and consignment, the environment can negatively affect the quality of the goods. In addition, the accountability for the goods is delegated from one party to another which makes a quality and quantity inspection necessary. The storage and distribution of complex event patterns as well as historical and real time event streams is not part of this paper. Research about this issue is being conducted and will be published soon. Here, we assume that the required event data will be provided by a sophisticated infrastructure.

The Service Task, as the central element, is placed at the pool of the retailer (see Figure 8) since the inspection belongs to his process activities.

As defined in our approach, all SELECT fields as well as FROM and WHERE attributes are not part of the graphical BPMN artifacts. For a better understanding, we nevertheless annotated some of the required attributes to the BPMN artifacts in Figure 8 and presented the attributes of the WHERE condition in table 5.

Table 5. Example for EPL-2-BPMN WHERE condition attribute

Attribute Name	Value
condition	temperature > -18 OR humidity > 0.8 OR manufacturing_date + 40 < current_timestamp()
cancelActivity	true

4   Related Work

So far, little attention has been devoted to research concerning usability aspects of CEP. Some related work to our approach can be found in the BPM research community. Here [10] provides a mapping of BPEL statements to BPMN artifacts. They integrate their mapping into the open-source modeling framework Oryx[1]. [11], on the other hand, discuss the mapping of BPMN to BPEL. Focusing on the same topic, [12] describes the mismatches occurring between the two languages. The difference to our work is that BPEL was exclusively designed to provide the execution of workflows, i.e., a mapping of BPEL and BPMN (and vice versa) is quite intuitive and straight forward. In contrast, EPL exclusively focuses on events, which makes it harder to define a complete mapping of all EPL statements to a process modeling language such as BPMN.

In [13], a new language for describing complex event patterns, called Business Event Modeling Notation (BEMN) is defined. The authors introduce a graphical notation and provide formal semantics. However, they do not explicitly address EPL, but provide a generic event modeling notation. A different approach by [14] presents a language independent process-modeling tool that is dynamically extensible and offers several export functionalities.

In the research area of agent systems [15] provides a first attempt of mapping business process diagrams to agent concepts through a graph based representation of BPMN. [16] presents an enterprise cockpit for performance monitoring of business processes by evaluating events coming from an "event cloud", while [17] focuses on event driven rules for faster responses to business critical events. They introduce a rule management system which enables users to graphically compose event-triggered rules for controlling the processing of services. In [18], the authors provide concepts and models for presenting and structuring events covering type inheritance and exheritance, dynamic type inferencing, attribute types, and extendability and addressability of events.

None of the before mentioned works directly focuses on the graphical representation of EPL, which indicates a major research gap our current and future work tries to close. In our opinion, the existing approaches for mapping complex event patterns are too specialized and are lacking the integration of existing processing technologies. Our approach focuses on an complex event modeling language that can be interpreted by existing CEP engines, which have already been applied in operative environments. Though we especially focused on the Esper EPL, other EPL language standards can easily be integrated, too.

---

[1] http://www.oryx-editor.org/

5   Discussion and Further Research

In this paper, we presented for each of the core EPL statements SELECT, FROM, and WHERE a graphical representation in BPMN 2.0 syntax. However, in order to provide a better usability for business users, our EPL-2-BPMN approach needs to be integrated into a graphical software tool, e.g., as a an Eclipse plugin or as part of the Oryx framework. Such a modeling tool should provide definitions and interfaces for further EPL constructs, such as complex WHERE conditions or aggregate functions on SELECT fields. In addition, the modeling tool should implement the basic framework as a single structure that can be extended through further data inputs of joined streams.

Since BPMN as well as EPL are non-formal languages, the proposed mapping of our approach cannot be evaluated by the use of formal methods. Nevertheless, we provided a profound representation based on the explicitly elaborated BPMN specification in order to facilitate a verification for semantic correctness.

We suggest research on the general categorization of real world events – a grouping of temperature, vibration, or acceleration events seems promising. Based on classified real world events, we also propose an extension of the currently available BPMN event artifacts. In particular, we suggest to derivate from the Conditional Event in order to create more specialized events. Corresponding graphical elements for these derivations would then significantly increase the usability and readability of event-driven business process models including EPL statements.

These further developments could lead to an EPL-2-BPMN cockpit showing the user the status of relevant processes, as well as deviations of planned processes across company borders. Additional interfaces to BPEL could provide possibilities for orchestration and choreography with other services.

6   Conclusion

In this paper, we presented an approach for integrating complex events, described in Esper EPL, into the modeling notation BPMN. In particular, we assigned each of the core EPL clauses – SELECT, FROM, and WHERE – to an corresponding BPMN artifact. Further, we provided a class diagram as basis for the development of a future EPL-2-BPMN modeling tool or EPL-2-BPMN Cockpit for process surveillance or flexible process adjustment. Through our work, we aim to achieve a better usability of CEP technologies fostering their further diffusion and adoption in the business domain.

Acknowledgments

This research was funded by the German Federal Ministry of Education and Research under grant number 01IA08001E as part of the Aletheia project[2]. The responsibility for this publication lies with the authors.

---

[2] http://www.aletheia-projekt.de/

References

1. Ammon, R., Silberbauer, C., Wolff, C.: Domain Specific Reference Models for Event Patterns – for Faster Developing of Business Activity Monitoring Applications. In: VIP Symposia on Internet Related Research with Elements of M+ I+ T++, vol. 16 (2007)
2. Luckham, D.: The Power of Events: An Introduction to Complex Event Processing in Distributed Enterprise Systems. Addison-Wesley Professional, Reading (May 2002)
3. Ibach, P., Bade, D., Kunz, S.: Smart Items in Ereignisgesteuerten Prozessketten. In: 39. Jahrestagung der Gesellschaft für Informatik (2009)
4. EsperTech: Esper - Reference Documentation (2009), http://esper.codehaus.org/esper-3.2.0/
5. Albek, E., Bax, E., Billock, G., Chandy, K.M., Swett, I.: An Event Processing Language (EPL) for Building Sense and Respond Applications. In: 19th IEEE International Parallel and Distributed Processing Symposium (Ipdps 2005), vol. 3 (2005)
6. Oracle: EPL Reference Guide (2009), http://download.oracle.com/docs/
7. Microsoft: .NET Developer Framework: LINQ, http://msdn.microsoft.com/en-us/netframework/aa904594.aspx
8. Object Management Group: Business Process Model and Notation (BPMN) 2.0. (November 2009), http://www.omg.org/spec/BPMN/2.0
9. EPCglobal: EPCglobal Tag Data Standards – Version 1.4. (August 2008), http://www.epcglobalinc.org/standards/
10. Weidlich, M., Decker, G., Großkopf, A., Weske, M.: BPEL to BPMN: the Myth of a Straight-Forward Mapping. In: Meersman, R., Tari, Z. (eds.) OTM 2008, Part I. LNCS, vol. 5331, pp. 265–282. Springer, Heidelberg (2008)
11. Ouyang, C., Dumas, M., Ter Hofstede, A., van der Aalst, W.: From BPMN Process Models to BPEL Web Services. In: ICWS, pp. 285–292. IEEE Computer Society, Los Alamitos (2006)
12. Recker, J., Mendling, J.: On the Translation between BPMN and BPEL: Conceptual Mismatch between Process Modeling Languages. In: Eleventh International Workshop on Exploring Modeling Methods in Systems Analysis and Design (EMMSAD 2006), pp. 521–532 (2006)
13. Barros, A., Decker, G., Grosskopf, A.: Complex Events in Business Processes. In: Abramowicz, W. (ed.) BIS 2007. LNCS, vol. 4439, pp. 29–40. Springer, Heidelberg (2007)
14. Küster, T., Heßler, A.: Towards Transformations from BPMN to Heterogeneous Systems. In: Mecella, M., Yang, J. (eds.) BPM 2008 Workshop Proceedings (2008)
15. Endert, H., Hirsch, B., Küster, T., Albayrak, S.: Towards a Mapping from BPMN to Agents. In: Huang, J., Kowalczyk, R., Maamar, Z., Martin, D., Müller, I., Stoutenburg, S., Sycara, K. (eds.) SOCASE 2007. LNCS, vol. 4504, pp. 92–106. Springer, Heidelberg (2007)
16. Jobst, D., Preissler, G.: Mapping Clouds of SOA-and Business-related Events for an Enterprise Cockpit in a Java-based Environment. In: Proceedings of the 4th International Symposium on Principles and Practice of Programming in Java, p. 236. ACM, New York (2006)
17. Schiefer, J., Rozsnyai, S., Rauscher, C., Saurer, G.: Event-driven Rules for Sensing and Responding to Business Situations. In: Proceedings of the 2007 Inaugural International Conference on Distributed Event-based Systems, pp. 198–205. ACM, New York (2007)
18. Rozsnyai, S., Schiefer, J., Schatten, A.: Concepts and Models for Typing Events for Event-based Systems. In: Proceedings of the 2007 Inaugural International Conference on Distributed Event-based Systems, p. 70. ACM, New York (2007)

Managing Business Process Variants at eBay

Emilian Pascalau[1] and Clemens Rath[2]

[1] Hasso Plattner Institute, University of Potsdam, Germany
emilian.pascalau@hpi.uni-potsdam.de
[2] European Center of Excellence, eBay, Ireland
crath@ebay.de

Abstract. The issue of business process variants management has been addressed several times already. However new situations with their own specific arise all the time, and proper solutions need to be developed in order to address such specific context. Process variants management at eBay is an example of such a situation. Variants are imposed by business facts, and only for a single process there could be more than 8000 variants. In this paper we introduce an ontology based approach to address the management of business process variants. The ontology based solution relates business context with reusable process flow elements, thus binding the reason for which a variant has been introduced with the variant itself. Variants can be later compared and queried for based on this relationship. In such a context, first, one has to model *for reuse* to be able to model *with reuse*. In order to provide a complete process variants management approach, the ontology based solution is complemented by an inheritance mechanism.

Keywords: business process variants, reuse, ontology, inheritance, process complexity.

1 Introduction

The recurrent issue of variants management in software engineering has proven to be a problem for business processes (see for example [1,2]). Despite the fact that several approaches have been developed to address the issue of process variants the subject is far from being resolved.

Management of process variants is also an issue for eBay. In this paper we introduce an ontology based approach to address the problem of process variants management. Our solution has been derived from a real industry scenario.

The eBay Customer Support provides individual services to buyers and sellers worldwide. Striving for the best possible individual customer service, the sheer number of customers, countries and languages served implies a huge demand for mass customization. Even though an individual service customization will be achieved case by case on an instance level; processes need to support the Customer Support Representatives (CSR) with tailored variants in the best possible way. As an example a Spanish business seller having an issue with mass listings would need a completely different support than an American casual customer

J. Mendling, M. Weidlich, and M. Weske (Eds.): BPMN 2010, LNBIP 67, pp. 91–105, 2010.

trying to sell his first item even both cases are dealing with the same question on how to list an item on eBay. Both cases would require different service levels, different entry points such as local phone numbers or web forms and of course different answers in different languages from different teams.

Process variants in the eBay context have different facets, e.g. a business function is performed differently based on a geographic location. The reason for variants is determined by many of the *business facts*, i.e. any entity that has a meaning for the business, i.e. laws, regulations, departments, customer types etc. Thus any element of the context it is called generically a *business fact*.

An ontology-based approach allows binding the reason for which a variant has been introduced with the variant itself. Having the reason that required the introduction of a new variant bound to the variant itself allows easy querying for variants and their comparison. We complement the ontology approach with an inheritance mechanism that contributes both to management and consistency of variants as well as structural comparison of variants.

This approach entails tackling of two perspectives: model *for reuse* and model *with reuse*.

Although the proposed solution can be used with any process modeling language we introduce it here using (Business Process Modeling Notation) BPMN[3] for two main reasons: eBay is using BPMN and BPMN is an Object Management Group (OMG) specification, already using an object model.

2   eBay Scenario

With more than 90 million active users globally, eBay is the world's largest online marketplace. eBay connects individual buyers and sellers, in 38 markets using 16 languages[1].

Customer segment	Sites	Payment	Department	CRM system	Channel
Casual Seller	BE-FR	PayPal	Safe Harbor BE	iPop	Mail
Top Buyer A/B	BE-NL	Bank Transfer	EU Fraud	Kana	Phone
Top Buyer C		Direct Debit	Top Customer Care	SAP CRM	Fax
Top Seller		Credit Card	Office of the President		Letter
Power Seller Bronce		CoD	Overflow Vendor X		Board
Power Seller Silver		...			Upload
Power Seller Gold					Walk in
Power Seller Platinum					

Fig. 1. excerpt of eBay taxonomies

---

[1] http://www.ebayinc.com/who; June 6th 2010.

Figure 1 depicts a small excerpt from eBay's taxonomies of business facts (i.e. Customer segment, CRM System). 6 taxonomies are grouped in Figure 1. A small subset of elements for each taxonomy is also present. The number of possible process variations is determined by the degree of freedom the system has, i.e., the number of possible arrangements of different business facts. The set of business facts depicted in Figure 1 influences a single eBay process. This particular business process is influenced by 6 business facts $\{b_1..b_6\}$, e.g., sites, payment, etc, that respectively have the following number of subtypes $\{8, 2, 5, 5, 3, 7\}$. This leads to a total number of 8400 possible variants.

Three of these possible variants of an eBay business process are depicted in Figure 2. These variants are imposed by different sites (countries, i.e. DE - Germany, UK - United Kingdom, NA - North America), different CRM systems and different Phone Systems.

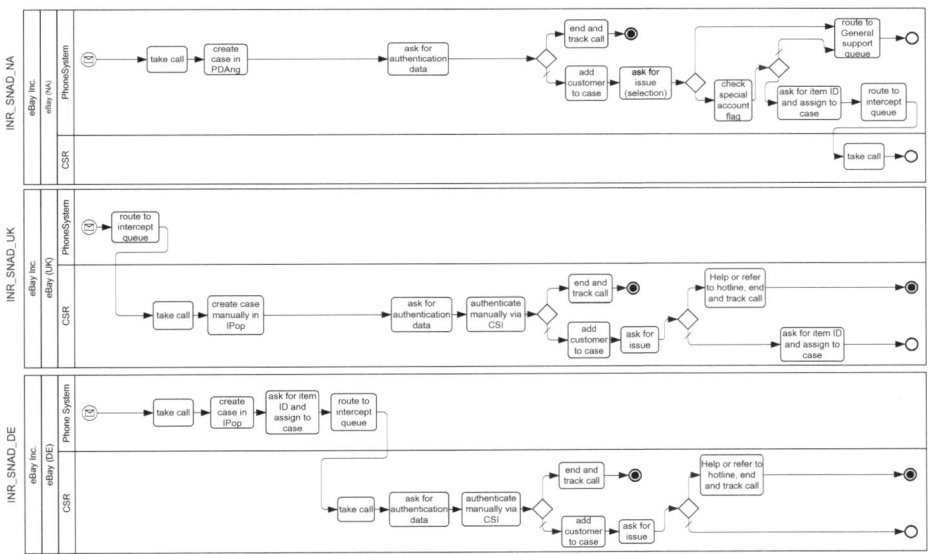

Fig. 2. Three eBay variants for a customer related process

When dealing with business processes a distinction is made between *process models* and *process instances*. Basically, as emphasized in [4] *a process model defines restrictions on process instances that belong to the process model* (see [4] for an extensive discussion). This distinction is quite important also in the management of process variants, as in some of the current approaches that address this issue (i.e. C-EPCs [1]), aEPCs [2] a *process variant* is defined as a *process model*, while for other approaches (e.g. [5,6]) a variant is defined as a *process instance*. In the context of the solution introduced here a *process variant* is defined as a *process model*.

3   eBay Scenario Discussion

The eBay scenario has several facets: causes of variations, identifications of variations, modeling and reuse of process elements.

3.1   Causes of Variation

Looking at the different reasons for process variants, two main reasons can be distinguished: Internal and external business facts.

External business facts can be considered as given. They cannot or just marginally be influenced by the corporate. These external causes comprise legal requirements, languages, differences in the banking system or shipping infrastructure as well as cultural differences such as willingness to cooperate or adequate tonality in customer communication.

However, internal business facts can be influenced and bear the potential for complexity reduction. Internal business facts comprise, e.g., infrastructure, team setups or customer segmentations. As an example, different phone or CRM systems can be consolidated. Some of the internal reasons are historically caused and so can be eliminated. The awareness of the differences of the existing variants will help to achieve this goal. The complexity reduction can lead to a more consistent customer experience and helps to achieve a qualitative growth and efficiency gains. But again it raises a need for determining differences.

In some cases, internal and external business facts are interdependent. As an example some Customer Support teams do have sufficient language skills to support customers in a written form via mail or chat whereas they might be challenged supporting all customers on the phone due to their accent or customers dialects. These facts also need to be considered in the relation of business facts.

3.2   Identification of Variation

As indicated above, there are several reasons why the business needs to be aware of the differences in process variants. Some of these differences are obvious such as languages or teams dealing with them and being accessible via specific phone numbers. Others are not that obvious and sometimes hidden in the systems, e.g. different thresholds for specific customer segments.

To identify these variants, processes can either be documented during process implementation and process changes or they can be found through process audits, Gemba[2] visits or process mining [7] based on available data produced.

Typically for the service industry there are many manual processes. As in all manual processes human beings take decisions and act in a way which is not always predictable and find workarounds to predefined workflows. This empowers CSRs to react flexibly on situational requirements and will in most cases help them to provide a reasonable customer service in cases where the standard procedure doesn't cover to the situation adequately. The disadvantage of this behavior is the limited reliability of the data gathered. This additional degree of freedom on an instance level limits the reliability of process mining in the service industry domain.

[2] http://en.wikipedia.org/wiki/Gemba

3.3   Modeling of Process Variants

Three different approaches on modeling process variants have been identified: *Clean sheet, Branching, Specifying.*

With the *clean sheet* approach a process variant gets redesigned from scratch. This approach usually gets applied in cases where the current process variants are not considered to be reasonable or where the sum of all business facts, i.e. the business environment, seems to be very different.

Branching means that an already existing process variant gets "copied" and modified to fit the business needs of the new variant. In this case usually a complete new process documentation gets created.

In both approaches, the *clean sheet* approach and the *branching* approach changes in the originating process variant will not affect the new process variants and vice versa. Any changes need either to be governed and aligned manually or will lead to an uncontrolled growth of complexity.

The third approach uses one single source for the process definition. Variants of this parent process refer to their parent process and any changes to the parent process will impact the child process variants. By doing so consistency between all variants can be achieved and process governance can be supported. This approach is closely related to inheritance on object oriented software development and should be further discussed.

3.4   Reuse of Process Elements

Reusing existing process parts provides many advantages: Based on experience with one variant continual improvements can be applied to all variants, existing infrastructure, software or content can be leveraged and most of all: the customer experience will remain consistent.

The reuse and recombination of modular process building blocks gives the opportunity to create flexible service processes out of existing variants. From a BPMN modeling perspective these building blocks can either be sub processes, tasks or even gateways and events. Each of these elements have attributes or methods to which inheritance can be applied to.

4   Business Process Variants Management Framework

Section 3 emphasized the fact that the process of dealing with process variants management requires to tackle several perspectives: the identification of variants, the modeling of, and the consistency maintenance between variants as well as the reuse of process elements. Our framework combines these aspects into one holistic approach in order to provide a solution to process variants management.

Although the benefits of reuse have been recognized for many years, only recently the transition towards reuse based development has been addressed [8]. However in order to support process element reuse, processes modeling must consider two facets: modeling *for* reuse and modeling *with* reuse. The two facets perspective has been already emphasized in the field of domain engineering [9].

Thus an effective process element reuse requires basically 2 major things: collections of process artifacts designed *for* reuse and a mechanism to retrieve, adapt, or create new ones. In such a way modeling *with* reuse is achieved. Thus our solution is centered on these two axes: *for reuse* and *with reuse*. For the *for reuse* axis (see Section 4.1) we define an ontological framework that comprises a set of high level concepts to semantically relate reusable elements with concepts from the business environment; concepts that define or impose variants for processes.

We address the *with reuse* axis using reasoning techniques complemented by an inheritance mechanism (Section 4.2).

4.1   Ontological Framework

Lately, context and context awareness has attracted a lot of attention (i.e. [10,11]). Coutaz et al. argues in [10] that *context is key* in the development of new services. Furthermore explicit relationship between environment and adaption are the key factors that will unlock context-aware computing at a global scale [10].

Dey in [11] defines context as *any information that can be used to characterize the situation of an entity. An entity is a person, place, or object that is considered relevant to the interaction between a user and an application, including the user and applications themselves.* Similarly for Coutaz et al. [10] the context is a *structured and unified view of the world in which the system operates.*

The influence of context to business processes has been already recognized [12].

In our scenario (see Section 2) the context is represented by all the elements which jointly create the business domain (i.e. customer types, departments, CSR systems, law requirements and such like). Explicit conceptualization of the context is out of the scope of this paper. For this issue one could refer to [12]. Recall that here, an element of the context it is called generically a *business fact*. With respect to the model vs. instance discussion, the context can influence both. Context can enforce the way processes are modeled and on the other hand the context can give *meaning* to a process instance and respectively to its underlying model. For example one could deal with a sel/buy process, a very generic one. But whenever contextual information is added, the *meaning* of a process could be totally different, as there is a big difference between selling tomatoes and selling i.e. chemical products.

However in terms of managing process variants (recall that here a process variant is a process model) context influences the modeling phase. The eBay scenario underlines two perspectives with respect to context: (1) context is the key factor that imposes the variants and (2) context is also the factor that will help in managing variants. The approach introduced here provides the ability to relate context elements with process flow elements. Thus searching for variants can be done using reasoning techniques.

As the context is a set of concepts and their relationships with each other, then a proper way to define it is by means of ontologies [13]. Uschold and Jasper argue in [14] that though an *ontology can take a variety of forms, it will include a vocabulary of terms, and some specification of their meaning.* Definition of concepts and their relationships impose a structure on the domain and constrain terms interpretation.

As in any other business and enterprise environment, the eBay context is based also on several ontologies, i.e. ontologies for CRM systems, for dealing with law enforcement and so forth. In addition these ontologies interfere with each other, as the domains that they model interfere with each other. Thus our ontological approach is based on the concept of an unified foundational ontology. For an in depth presentation and corresponding formalization, one should refer to [15].

Ontologies are characterized by several benefits [14], but some of them are of particular interest to the problem of business process variants management. We adapt from [14] a set of characteristics that help in tackling the variants management issue in the eBay context.

- Reusability: ontologies as previously argued are the basis for the formal capturing of business entities, attributes and processes, and their relationships. Thus the formal representation can be used as a reusable and/or shared artifact
- Search: ontologies can be used as meta-data in searching repositories of reusable artifacts or process variants.
- Specification: can assist in the process of designing and modeling of process variants. In the eBay context ontological concepts are those that impose variations to processes. Using the relationship between business facts and reusable process flow elements, the latter ones can be easily identified in the repository. Thus the specification design process is improved.
- Maintenance and governance: as ontologies serve to improve documentation, understanding, and as already argued they even impose variants, they also help in reducing maintenance of process variants. In general maintenance is achieved by using an ontology as a neutral authoring language for different target languages. However for our approach using an ontology improves maintenance of variants, as variants are related to the ontological concepts. Thus searching for a specific process variant is improved; e.g. the modeler does not have to search for a process variant based on a process id, but it does this by searching processes that are related to one or a set of business facts.

Figure 3 depicts part of our ontological framework in a MOF/UML (Meta-Object Facility / Unified Modeling Language). On one side it depicts an excerpt from the BPMN 2.0 [3] specification and on the other hand the new concepts that we introduce. The new concepts that we introduced are drawn either with a gray background or a gray edge. We are using UML (Unified Modeling Language) [16] to formalize our ontological framework as BPMN 2.0 specification is using it, and in addition UML being considered as *de facto* standard modeling language [15].

Recall that ontologies capture concepts and their relationships with each other basically in the form of a vocabulary of terms with their intended meaning. Figure 3 comprises terms for the concepts of `Process` and the valid concepts that form a process (`Activities`, `Gateways`, `Events`, `SequenceFlows` and so forth). Any of the `FlowNodes` is understood to be a reusable artifact or building

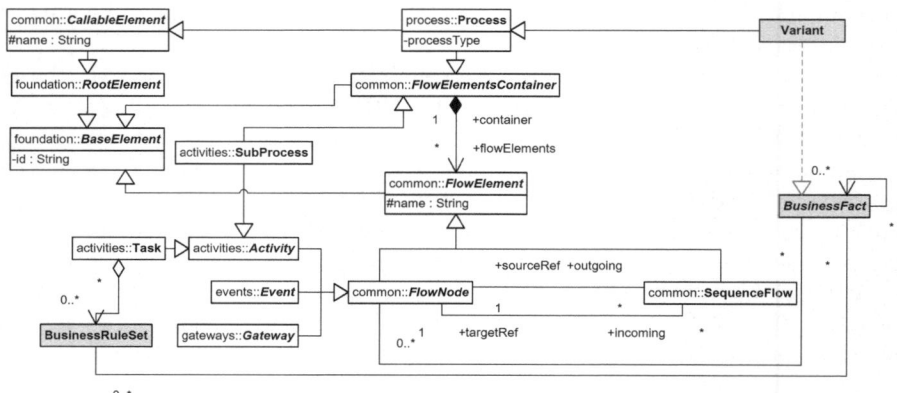

Fig. 3. Ontological framework

block. The more specific terms that make a FlowNode can be modeled *for reuse*. In terms of our framework to model *for reuse* means that a relationship exists between a FlowNode and a BusinessFact. In addition the reusable FlowNode it is stored in the repository together with its BusinessFact relationship. Note that the relationship is bidirectional, this implies that the relationship can be navigated from both sides (i.e. one can search for a FlowNode and already know which are the related BusinessFacts, or viceversa). A BusinessFact is a central concept of our framework. It subsumes any business concept. Examples of such business concepts are grouped in Figure 1.

Note that a Task could be implemented by means of rule sets (BusinessRuleSet) Rule sets are not part of the BPMN 2.0 specification either. However they are also imposed by the context, and thus there is an association with a BusinessFact. Relationships between BusinessFacts can also exist as described in the country-specific meta context in Section 2.

According to the ontology a Variant is a Process. The relationship between BusinessFacts and Processes exists indirectly, as a Process is a collection of FlowElements. The relationship BusinessFact - FlowNode allows a higher granularity and flexibility in dealing with variants.

The organizational view is also important for our variants management framework. Based on the example depicted in Figure 2 we can see that different FlowNodes belong to a different lane in different variants. Since the BPMN 2.0 specification it is unclear with respect to the definition of Pools and Lanes we provide our view on the matter in Figure 4. The previously described relationship between BusinessFacts and FlowNodes already tackles the organizational perspective, as there is a relationship between FlowElements and Lanes.

Figure 4 depicts the organizational view from the business processes's perspective (pools and lanes). However for the ontological view we stick to the one introduced in [4]. In addition we have to emphasize that a BusinessFact can represent also an organizational concept. Thus i.e. eBay's billing department, or a CRM system as in Figure 2.

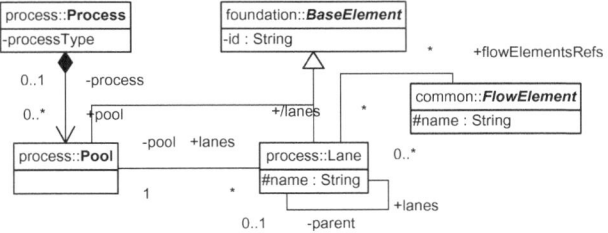

Fig. 4. BPMN meta-model excerpt - Pools and Lanes

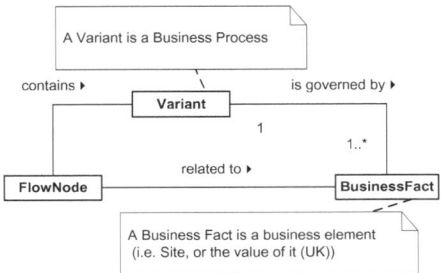

Fig. 5. Framework made simple

Figure 5 depicts a simpler view of the framework. Any `FlowNode` can be related to one or more `BusinessFacts`. A `Variant` which is a business process contains `FlowNodes`. `Variants` are imposed by at least one or more `BusinessFact`.

4.2   Framework's Modeling *with Reuse* Mechanism

Having discussed the modeling *for reuse*, by means of an ontological framework, we need a mechanism for the modeling *with reuse*; a mechanism that has to blend with the ontological framework introduced in 4.1.

Inheritance is a well known mechanism from software engineering that allows to derive implementation from a superclass [17]. The general view is that inheritance is a mechanism for incremental programming [18]. This characteristic of incremental programming makes the inheritance mechanism an appealing opportunity for our process management solution.

eBay context (see Section 2) actually requires such an approach, as it would be almost impossible to define 8000 variants for a single process in a different manner or i.e. using an approach based on configurable process models [1]. Using a configurable process model that encapsulates all the variants supposes that the modeler is aware of all the possible configurations. In addition each time a configuration has to be extracted, the modeler is required to scroll through the whole model and adjust every configurable node.

On the other hand using an inheritance based approach the whole process of creation and management of variants is easier, as it starts with a basic process, which is adjusted based on needs. Moreover the foreseen inheritance mechanisms allow not only to inherit/modify/add behavior with respect to reusable artifacts (`FlowNodes`) but by means of *mixin* inheritance, complete parts of processes can be mixed in to create a new variant. The term *mixin* is adopted from software engineering.

The 3 eBay variants depicted in Figure 2 underline already the necessity for several types of inheritance: *inheritance of flow, inheritance of organizational aspects*, as well as *inheritance by restriction*. It is easy to see that the 3 variants have common flow (i.e. "ask for authentication data", "authenticate manually via CSI", "end and track call", "add customer to case" etc.). There are common activities that belong to a different lane (e.g. "take call"), or activities that do not exist in all the lanes ("ask for item ID and assign to case").

The following inheritance types can be applied to process inheritance:

- *basic flow inheritance*: Deals with inheritance of basic flow. Basic flow is represented by `FlowElements` (see Figure 3).
- *organizational inheritance*: Concerns inheritance of organizational attributes of a business process. `Pools` and `Lanes` are taken into account and `FlowElements` that are related to them.
- *inheritance by restriction*: When e.g. some flow artifacts are not required, they have to be restricted from being inherited.
- *multiple inheritance*: inheritance form multiple processes. It is based on the previous types of inheritance.
- *mixin based inheritance*: This type of inheritance concerns inheritance of parts of business processes by combination.
- *inheritance from business facts*: Figure 3 contains a special type of inheritance, depicted in gray and with a dashed line. It implies that based on the `BusinessFact` - `FlowNode` relationship. A variant inherits from a `BusinessFact` via this relationship all the associated `FlowNodes`.

Figure 6 depicts the basic inheritance mechanism: *flow inheritance*. Note that we use the same terminology as in software engineering to address *subclassprocesses* and *superclassprocesses*. In software engineering the class from which behavior is inherited is called a superclass. The class that inherits behavior is called a subclass. Thus we use *superclassprocess* to denote the process from which we inherit behavior and *subclassprocess* to denote the more specific process, the one that inherits behavior.

The process entitled `superclassProcess` in Figure 6 has been defined based on three process variants from Figure 2. All common flow elements of the three variants make the *superclassProcess*, or the greatest common divisor with respect to flow elements of the three eBay variants in Figure 2.

The second process in Figure 6 entitled `subclassProcess` (eBay UK) is the UK variant from Figure 2. Both the *superclassprocess* and the *subclassprocess* contain only the flow elements, without organizational information. All the differences between the *superclass* and the *subclassprocess* are depicted in gray.

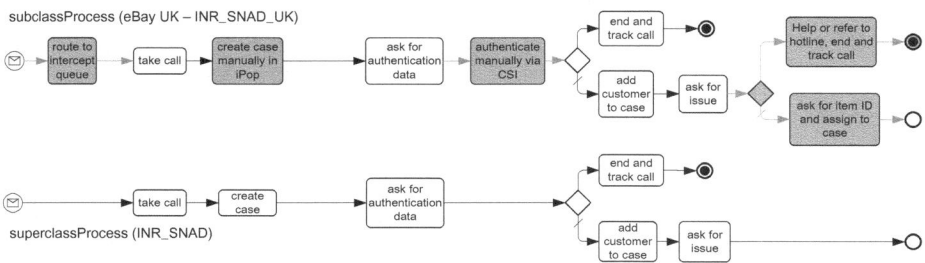

Fig. 6. Basic flow inheritance

As opposed to other approaches i.e. [1] here the ontological framework allows to define dynamic behavior with static constructs. Thus, e.g. a sequence edge is represented by the `SequenceFlow` concept (see Figure 3), a simple task by the `Task` concept and so forth.

Although similar, the inheritance mechanism, known in programming languages such as Java [17], cannot just be reused in the business processes inheritance, as described here. Recall that all elements composed in a business process are represented according to the ontological framework in Section 4.1 as static constructs, *type* based. Thus e.g. to override the sequence flow that connects the `start message event` and `take call` in `superclassProcess` (Figure 6) with a sequence flow that connects the `start message event` and `route to intercept queue` in `subclassProcess` requires that only the value of the attribute `targetRef` has to be updated. However such a task cannot be performed using structure inheritance in a programming language such as Java.

To model the `subclassProcess` from Figure 6 via inheritance we need a set of transition rules as in software engineering: *override, add* etc. The sequence flow that connects the `start message event` from the `superclassProcess` and the `take call` task are changed in the subclassprocess with an *override* transition and two *adds*. In the `subclassProcess` we have the overridden sequence flow that now connects the `start message event` and the newly added `route to intercept queue` task and also a newly added sequence flow that connects the `route to intercept queue` and the `take call` task. A programming language such as Java does provide an override transition but only for methods and not for properties as our ontological framework requires. Thus business process inheritance is similar but not the same and inheritance mechanisms from programming languages such as Java cannot be directly translated to business processes.

5   Usage Example

Figure 7 presents a small example on how the ontology proposed here can be used to manage business process variants. `Site` and `CRMSystem` are `BusinessFacts`. Each of these `BusinessFacts` is further specialized in `UK` and `NA` for `Site` and `IPop` and `PDAng` for `CRMSystem`. `CreateCaseManuallyInIPop` and `CreateCaseinPDAng` are `Tasks`. `CreateCaseManuallyInIPop` is related to the `UK`

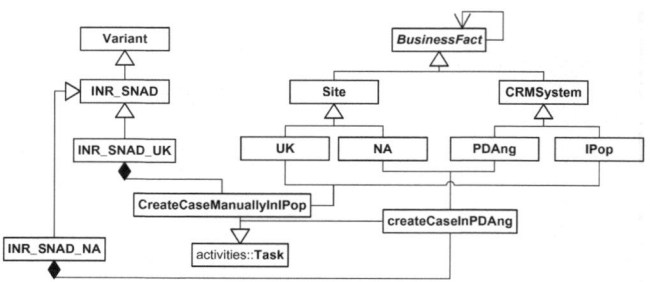

Fig. 7. Ontology usage example

and IPop business facts. On the other hand CreateCaseinPDAng is in a relationship with PDAng and NA. INR_SNAD_UK (Figure 2 and Figure 6) and INR_SNAD_NA (Figure 2) are both variants of the same process INR_SNAD (Figure 6).

Based on the example depicted in Figure 7 we immediately find out what the differences between INR_SNAD_UK and INR_SNAD_NA variants are: different sites and CRM systems, as well as different tasks.

Finding processes that are related to UK and IPop business facts is also simple: INR_SNAD_UK.

The ontological approach proposed here can be put in practice using ontology and rule-based reasoning techniques and tools (i.e. using Protege[3], Drools[4]).

6   Related Work

Variants management is an issue addressed in various domains ranging from software configuration management [19,20,21] and feature diagrams [22,23,24] to variants of process models nowadays. Several approaches have been defined for process models as well.

C-EPCs [1] allows the configuration of process models by distinguishing between choices that can be made at runtime and those that have to be made before, i.e., configuration time. Via *configurable nodes* EPCs can be configured. On the other hand aEPC [2] works on the principle of *projection* [25] and only elements that have a particular label are included in the extracted model. Inheritance of behavior in workflows [26,27] is a formal approach for tackling problems that are related to change. Four inheritance rules (*protocol inheritance, projection inheritance, protocol/projection inheritance, life-cycle inheritance*) are defined to tackle dynamic change.

In comparison to C-EPC and aEPC that have a holistic model as starting point, our approach does not require the existence of such a model. Moreover in the eBay context creation of such a holistic process would be almost impossible, given the fact that only for a single process, more than 8000 variants have to

---

[3] http://protege.stanford.edu/
[4] http://jboss.org/drools

be modeled. The difference between variants can be extracted with respect to business facts.

La Rosa [28] defines a questionnaire approach on top of C-EPCs to improve management of variants. Though a C-EPC model is still required, and based on it a domain specific questionnaire model is defined. However to empower a user to extract a configured model, in addition to the reference model and the questionnaire model a mapping between these two models is also required.

Existing inheritance [26,27] based approaches do not address the relationship with business facts also. Thus differences between variants can be computed only from a process flows perspective.

Annotations based approaches, e.g., the PESOA (Process Family Engineering in Service-Oriented Applications) project [29,30] defines so-called *variant-rich process models* as process models that are extended with stereotype annotations to accommodate variability. Both UML Activity Diagrams as well as BPMN models can be tackled with this approach. The places within a process where variability can occur are marked with the stereotype `VarPoint`. Several other stereo types , e.g., `Variant`, `Default`, `Abstract`, `Alternative`, `Null`, `Optional`) are used to specify different configuration options. Compared to our approach, we do not predetermine configuration points.

Others [5,6] address variants at execution time. The notion of process constraints it is used to tackle the need for flexibility and dynamic change [31] at execution time. However in [5,6] a process instance is taken for a process variant.

7   Conclusion

Managing business process variants at eBay is a challenging task as the business context is the one that imposes variants. A single process in the eBay context could have more than 8000 variants. Business facts such as countries, specific laws for each country or different technical systems are only a few examples of business facts that would generate a process variant. The differences between variants are both semantical and structural. Semantical because of the different business facts which trace out the variant and structural as new reusable process artifacts are overridden, added, restricted etc.

The approach that we introduced in this paper starts from the business context (the whole pallet of business facts) which is already existent at eBay. Though the business context is the one that imposes the variants, we use it to our advantage to manage variants. We have introduced an ontological approach that relates business facts with reusable process elements. Thus the reason for which a variant has been introduced is stored in the system. This simple relationship is also a key factor in retrieving the right processes, as well as finding the correct reusable process elements that can be later used in defining new processes.

Two management perspectives are addressed by our approach: (1) modeling *for reuse* and (2) modeling *with reuse*.

The semantic ontological framework it is complemented by an inheritance mechanism, that deals with the structural differences. The inheritance mechanism provides the execution mechanism, the set of rules which together with the ontological framework provide the means to define and maintain variants. Relationship between variants' structural modification is also maintained. Also through the inheritance mechanism, consistency between variants it is enforced. The inheritance has an additional benefit, as maintenance of variants is improved. Changes made on *superclassprocesses* in the inheritance chain will be automatically translated to any *subclassprocesses* that belong to the inheritance chain.

Future research will concentrate on defining concrete algorithms for the different types of inheritance introduced here. In addition to that eBay is willing to engage in prototyping the framework.

Albeit the solution introduced here is based on the BPMN meta-model, the fundamental ideas behind it can be easily translated to other process modeling languages i.e. EPCs, YAWL, if they are formally defined by a meta-model. The reusable concepts have to be related to business facts. The inheritance process in principle should work out of the box, but based on the modeling language slight adjustments might be required.

References

1. Rosemann, M., van der Aalst, W.M.P.: A configurable reference modeling language. Inf. Sys. 32(1) (2007)
2. Reijers, H., Mans, R., van der Toorn, R.: Improved model management with aggregated business process models. DKE 68(2) (2009)
3. OMG: Business Process Model and Notation (BPMN). FTF Beta 1 for Version 2.0. (August 2009), http://www.omg.org/spec/BPMN/2.0
4. Weske, M.: Business Process Management: Concepts, Languages, Architectures. Springer, Heidelberg (2007)
5. Lu, R., Sadiq, S., Governatori, G., Yang, X.: Defining Adaptation Constraints for Business Process Variants. In: Proceedings of the 12th International Conference on Business Information Systems (BIS 2009), pp. 145–156 (2009)
6. Lu, R., Sadiq, S., Governatori, G.: On Managing Business Processes Variants. Data & Knowledge Engineering 68(7), 642–664 (2009)
7. van der Aalst, W., Weijters, A., Maruster, L.: Workflow Mining: Discovering Process Models from Event Logs. IEEE Transactions on Knowledge and Data Engineering 16(9), 1128–1142 (2004)
8. Sommerville, I.: Software Engineering, vol. 8. Addison Wesley, Reading (2007)
9. de Almeida Falbo, R., Guizzardi, G., Duarte, K.C., Natali, A.C.C.: Developing Software for and with Reuse: An Ontological Approach. In: Proceedings of the ACIS International Conference on Computer Science, Software Engineering, Information Technology, e-Business and Applications (CSITeA 2002), Foz do Iguacu, Brazil (2002)
10. Coutaz, J., Crowley, J.L., Dobson, S., Garlan, D.: Context is key. Communications of the ACM 48(3), 49–53 (2005)
11. Dey, A.K., Abowd, G.D.: Towards a better understanding of context and context-awareness. Technical Report GIT-GVU-99-22, Georgia Institute of Technology (1999)
12. Rosemann, M., Recker, J., Flender, C.: Contextualisation of Business Processes. International Journal of Business Process Integration and Management 3(1), 47–60 (2008)

13. Guarino, N.: Formal Ontology and Information Systems. In: Proceedings of FOIS 1998, Trento, Italy, June 6-8, pp. 3–15. IOS Press, Amsterdam (1998)
14. Uschold, M., Jasper, R.: A Framework for Understanding and Classifying Ontology Applications. In: Proceedings of the IJCAI 1999 workshop on Ontologies and Problem-Solving Methods (KRR5), Stockholm, Sweden (1999)
15. Guizzardi, G.: Ontological Foundations for Structural Conceptual Models. PhD thesis, Telematics Instituut, Enschede, The Netherlands, Telematica Instituut Fundamental Research Series No. 15, ISBN 90-75176-81-3 (2005)
16. OMG: UML 2.0 superstructure specification (2005), http://www.omg.org/spec/UML/2.0/Superstructure/PDF/
17. Gosling, J., Joy, B., Steele, G., Bracha, G.: The Java Language Specification, 3rd edn. Addison-Wesley, Reading (2005)
18. Bracha, G., Cook, W.: Mixin-based inheritance. In: Proceedings of the European Conference on Object Oriented Programming Systems Languages and Applications (OOPSLA 1990), pp. 303–311. ACM, New York (1990)
19. Estublier, J., Casallas, R.: The Adele Software Configuration Manager. In: Configuration Management. Trends in Software. John Wiley & Sons, Chichester (1994)
20. Tryggeseth, E., Gulla, B., Conradi, R.: Modelling Systems with Variability using the PROTEUS Configuration Language. In: Proceedings of the Int. Conference on Software Configuration Management (1995)
21. Turkay, E., Gokhale, A., Natarajan, B.: Addressing the Middleware Configuration Challenges using Model-based Techniques. In: Proceedings of the 42nd Annual ACM Southeast Regional Conference (2004)
22. Batory, D., Geraci, B.: Composition Validation and Subjectivity in GenVoca Generators. IEEE Transactions on Software Engineering 23(2) (1997)
23. Czarnecki, K., Helsen, S., Eisenecker, U.: Formalizing Cardinality-Based Feature Models and Their Specialization. Software Process: Improvement and Practice 10(1) (2005)
24. Schobbens, P., Heymans, P., Trigaux, J.: Feature Diagrams: A Survey and a Formal Semantics. In: RE (2006)
25. Baier, T., Pascalau, E., Mendling, J.: On the Suitability of Aggregated and Configurable Business Process Models. In: Nurcan, S., Ukor, R. (eds.) BPMDS 2010 and EMMSAD 2010. LNBIP, vol. 50, pp. 108–119. Springer, Heidelberg (2010)
26. van der Aalst, W., Basten, T.: Inheritance of Workflows: An approach to tackling problems related to change. Computing Science Report 99(6) (1999)
27. Basten, T., van der Aalst, W.: Inheritance of Behavior. Computing Science Report 99(17) (1999)
28. Rosa, M.L.: Managing Variability in Process-Aware Information Systems. PhD thesis, Queensland University of Technology, Brisbane, Australia (2009)
29. Puhlmann, F., Schnieders, A., Weiland, J., Weske, M.: Variability mechanisms for process models. Technical Report 17/2005, Hasso-Plattner-Institut (June 2005)
30. Schnieders, A., Puhlmann, F.: Variability Mechanisms in E-Business Process Families. In: Proceddings of the International Conference on Business Information Systems (BIS 2006), LNI, vol. 85, pp. 583–601. GI (2006)
31. Dadam, P., Reichert, M.: The ADEPT project: a decade of research and development for robust and flexible process support. Challenges and Achievements. Computer Science - Research and Development 23(2), 81–97 (2009)

Managing Variability in Process Models by Structural Decomposition

Maria Rastrepkina

Business Process Technology Group
Hasso Plattner Institute at the University of Potsdam
Prof.-Dr.-Helmert-Str. 2–3, D-14482 Potsdam, Germany
Maria.Rastrepkina@student.hpi.uni-potsdam.de

Abstract. Business process management (BPM) provides companies with techniques for analyzing, designing, and executing operational processes which are documented as process models. Often, companies face the problem of supporting similar processes, i.e., process variants, within their organizational boundaries. This may be due to cultural particularities of running business or/and different legal regulations enforced at different departments operating worldwide. Effective mechanisms for managing process variants should allow companies to achieve operational consistency and effective reuse of their process repositories. This paper presents a novel approach for managing variability in process models based on the structural decomposition and representation of process variants as an integrated set of building blocks called process modules.

Keywords: Process management, process variants, variability management, process reuse, process modularization.

1 Introduction

Effective management of business processes allows companies to decrease their costs and time to market significantly [13]. To satisfy all domain requirements, different variations within a business process may occur resulting in a multitude of similar business processes, i.e., process variants [4,6]. For instance, Fig.1(a) and (b) show variants of the order-to-cash business process of divisions A and B of a reseller company. These process variants have arisen due to the fact that division A requires that a payment is received before a product has been sent, and division B allows to respite a payment.

Process variants can be captured as separate process models, as represented in Fig.1. However, this strategy has some drawbacks. First, it leads to redundancy. Second, in case of any change which has to be applied to all process variants, each process model has to be adjusted separately. Third, separation of process models restricts the potential for reuse of existing solutions and for combining process variants into new ones. The factors listed above aggravate management of process variants significantly turning it into a time-consuming and error-prone task.

J. Mendling, M. Weidlich, and M. Weske (Eds.): BPMN 2010, LNBIP 67, pp. 106–113, 2010.

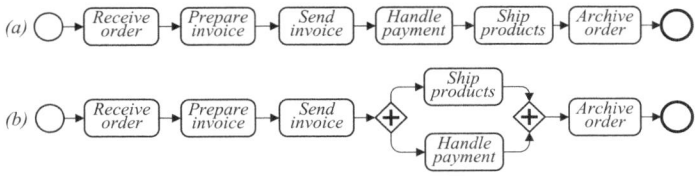

Fig. 1. Order-to-cash process variants of a reseller company

In this paper we present an approach which on the one hand aims at managing variability of process models, and on the other hand supports a user during design and customization of process variants. Within a proposed approach a family of process variants is represented as an integrated set of building blocks called *process modules*. Each process module represents a fine-grained part of a process model and captures a common aspect or a difference between process variants. Process modules can be assembled into an individual process variant by *configuration*.

This paper is structured as follows. Section 2 gives an outlook of the related work. Section 3 summarizes a number of auxiliary terms. Section 4 presents a technique for managing process model variants. Finally, Section 5 concludes the paper and discusses the future work.

2   Related Work

BPM community proposed several techniques for capturing variability in process models. Rosemann and van der Aalst in [10] introduced the concept of *configurable process model* and presented the Configurable Event-driven Process Chains (C-EPCs) for variability modeling. C-EPC extends the EPC modeling language with a concept of configurable functions which can be configured as included, excluded, or conditionally skipped, and configurable connectors which can be mapped to equally or less restrictive connectors. In [2], the authors have extended the YAWL with input and output ports to capture configuration points. Through a configuration each task of C-YAWL model can be enabled, blocked, or skipped by configuring its ports. Becker et al. in [1] introduced an approach to hide element types in EPCs for configuration purposes. Within this approach, a *configurative reference model* is defined as a single integrated model that contains specific information for all company characteristics, and the specific models are provided by creating *views*. Puhlmann et al. [11] proposed to represent a variability of a software system in the process model. They introduced a concept of a *variant-rich process model*, which is a process model extended with stereotype annotations. During a configuration, each variation point is assigned with one or more variants. In [6], Hallerbach et al. present an operational approach called Provop. Within Provop a configurable process model is represented as a *generic model*, and configuration points are represented as adjustment points. Process variants are defined by a set of change operations, which are called options.

The presented approaches focus on the solving the problem of how to model variability and present different techniques to capture variation points in a process model. However, the questions of construction and customization of a configurable process model, e.g., the addition of a new process variant, have not been sufficiently addressed within the discussed works. As a solution, one can employ different techniques from the field of graph matching and merging [3], or process mining techniques [5]. Still, as it has been argued in [4], *"the creation of the configurable models required significant efforts"*. Another open question is how the techniques listed above can facilitate a customization of configurable process models, since *"it is risky and costly"* [10].

3 Preliminaries

In this section we provide a number of auxiliary concepts. We start with the definition of a process graph, adopted from [13].

Definition 1 (Process Graph). A tuple $P = (V, E, type)$ is a *process graph*, where: $V = T_P \cup G_P \cup \{i_P\} \cup \{o_P\}$ is a set of nodes in P where T_P is a set of tasks, G_P is a set of gateways, i_P - is an input node, o_P is an output node; $E \subseteq V \times V$ is a set of directed edges between nodes representing control flow such that (V, E) is a connected graph; $type : G_P \rightarrow \{and, xor, or\}$ is a function that assigns to each gateway a control flow construct.

In addition, we say that for a node $v \in V$, $\bullet v := \{y \in V | (y, v) \in E\}$, $v \bullet := \{y \in V | (v, y) \in E\}$. Let L be a universal set of labels, $L_P \subseteq L$ is a subset of labels, such that there exists a function *label* that assigns to each task of a process graph P a label. We require a process graph to be structurally correct, i.e. it must fulfill a set of requirements. We say that $\bullet i = \emptyset$, $o \bullet = \emptyset$, $\forall t \in T : |\bullet t| = 1 \wedge |t \bullet| = 1$. Each gateway is either split or join. A gateway $g \in G$ is a split if $(|\bullet g| = 1 \wedge |g \bullet| > 1)$, a gateway $g \in G$ is a join if $(|\bullet g| > 1 \wedge |g \bullet| = 1)$. Each process model node lies on a path from an input node to an output node. In the remainder, we do not specify execution semantics of a process graph. We assume that an interpretation of the process graph follows on common execution semantics presented in the existing works, and that a process graph is semantically correct (as example see [7]). In the following, we define a process subgraph and give a classification of its nodes, based on the concepts presented in [9,12].

A tuple $s = (V', E', type')$ is a *process subgraph* of a process graph $P = (V, E, type)$ if $V' \subseteq V$, $E' = E \cap (V' \times V')$, $type' = type|_{V'}$ where $type'$ is a restriction of function $type$ of P to the set V'. For a node $v \in V$, $in(v) = \{(y, v) | y \in \bullet v\}$ is a set of *incoming edges* of v, $out(v) = \{(v, y) | y \in v \bullet\}$ is a set of *outgoing edges* of v. E^+ denotes a transitive closure of E. A node $v \in V'$ of a process subgraph s is an *interior node* of s if $(in(v) \cup out(v)) \subseteq E'$, otherwise v is a *boundary node* of s. If $v \in V'$ is a boundary node, it is an *entry* of s if $in(v) \cap E' = \emptyset \vee out(v) \subseteq E'$, or an *exit* of s, if $out(v) \cap E' = \emptyset \vee in(v) \subseteq E'$. A process subgraph f is a *process fragment*, if it has exactly two boundary nodes, one is an entry node, denoted by $\triangleright f$, and one is an exit node, denoted by $f \triangleright$, and $\forall v \in V'$: $(\triangleright f, v) \in E^+ \wedge (v, f \triangleright) \in E^+$. For any process fragments q, w with a set of edges $E_q \subseteq E$ and $E_w \subseteq E$, respectively, we say that q, w are *nested* if $E_q \subseteq E_w$ or $E_w \subseteq E_q$. q, w are *disjoined* if $E_q \cap E_w = \emptyset$. If q, w are neither nested nor disjoined, we say they *overlap*.

Fig.2 exemplifies the structurally correct process graph which corresponds to the process variant from Fig.1 (b) with depicted process fragments. Process fragments can be obtained by performing structural decomposition. A unique and modular decomposition which is called the Refined Process Structure Tree (RPST) was proposed by Vanhatalo et al. in [12]. The RPST is a containment hierarchy of canonical fragments of a process graph, where the canonical fragments are exactly those fragments that do not overlap with any fragment in a set of process fragments of a process graph. We utilize the RPST within our approach, since it is finer grained than any known alternative approach and it can be computed in time linear to the number of edges of the graph [9]. All non-canonical

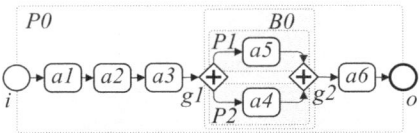

Fig. 2. Example of process graph with depicted process fragments

process fragments can be derived from the RPST from the canonical fragments as unions of their child fragments [12].

4 Module-Based Variability Management

In this section we introduce a concept of process modules, give a set of formal definition, and describe how process modules can be used to capture process variants.

4.1 Process Modules and Their Relations

Structural decomposition of process graphs allows to represent a process graph as a hierarchy of process fragments. Process graphs which capture variants of one business process share common fragments, and characterized by variable fragments, i.e. by fragments that differ. Variable fragments can be seen as *variation points* in a process model, so that any process fragment in such a point can be replaced by another process fragment. To be able to represent an alternative aggregation of process fragments we introduce a new structure – a process module. A process module specifies the structure of a given process fragment, it is not concrete but contains a set of *placeholders*. Each placeholder of a process module can be filled by another process module, i.e. *instantiated*. We introduce five types of process modules: trivial, polygon, bond, rigid, and variation point modules (see Fig.3).

Definition 2 (Process Modules). Let $P = (V, E, type)$ be a process graph with a set of nodes $V = T_P \cup G_P \cup \{i_P\} \cup \{o_P\}$, and Γ be a universal set of placeholders.

- A *trivial module* is a tuple $m = (u, ttype, tmodtype)$, where: $u \in V$ is a node in P, $ttype = type|_{\{u\}}$ is a restriction of function $type$ of P to the set $\{u\}$, $tmodtype \in \{input, output, task, gateway, silent\}$ is a type of a trivial module.

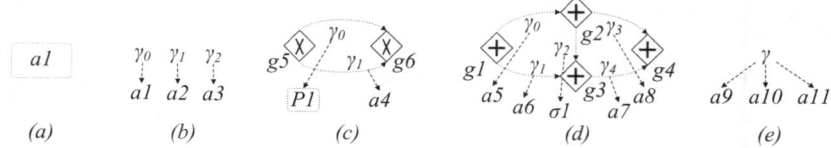

Fig. 3. (a) Trivial, (b) polygon, (c) bond, (d) rigid, (e) variation point modules

- A *polygon module* is a tuple $m = (\Gamma_m, order)$, where: $\Gamma_m \subseteq \Gamma$ is a set of placeholders, $order : \Gamma_m \to \mathbb{N}$ is an ordering function which assign an order position to a placeholders. A placeholder $\gamma \in \Gamma_m$ with an order position $k \in \mathbb{N}$ is denoted as γ_k.
- A *bond module* is a tuple $m = (\Gamma_m, entry, exit, btype, bplac)$, where: $\Gamma_m \subseteq \Gamma$ is a set of placeholders, $entry, exit \in G_P$ are gateways in P, which are called an *entry* and an *exit* nodes of m, respectively, $btype = type|_{\{entry, exit\}}$ is a restriction of function $type$ of P to the set $\{entry, exit\}$, $bplac : \Gamma_m \to \{(entry, exit), (exit, entry)\}$ is a function which assigns to a placeholder either an edge from an entry node to an exit node m, or an edge from an exit node to an entry node of m.
- A *rigid module* is a tuple $m = (\Gamma_m, RG, ree, rtype, rplac)$, where: $\Gamma_m \subseteq \Gamma$ is a set of placeholders, $RG \subseteq G_P$ is a set of gateways in P, $ree : \{g_1, g_2\} \in RG \to \{entry, exit\}$ is a function, which assigns a type to two gateways g_1, $g_2 \in RG$, such thay $ree(g_1) \neq ree(g_2)$, $rtype = type|_{RG}$ is a restriction of function $type$ of P to the set RG, $rplac : \Gamma_m \to RG \times RG$ is a function which assigns to a placeholder an edge which connects nodes of RG.
- A *variation point module* $m = \Gamma_m$, such that $|\Gamma_m| = 1$.

One can assign process modules to placeholders, see Fig.3. Process modules assigned to placeholders of a process module are *child modules*, and the process module is a *parent module*. A placeholder of parent module can be instantiated by a child module which is assigned to it. The relations between a parent module and its child modules are categorized as AND-, XOR-, and OR-relations. If process modules are in *AND-relation*, then only one child process module can be assigned to each placeholder of a parent module. Hence, parent module aggregates all its child process modules. XOR- and OR-relations indicate the alternative between child process modules. If process modules are in *XOR-relation*, then several child process modules can be assigned to a placeholder of a parent module. Only one child module can instantiate that placeholder. If process modules are in *OR-relation*, then several child process modules can be assigned to a placeholder of a parent module. Some of these child modules can instantiate that placeholder. Thus, in case of XOR- and OR-relations a choice between alternative child process modules has to be done before instantiation.

A set of process modules with their relations forms the *Configurable Module-based Process Structure Graph* (C-MPSG). It is a graph where trivial modules are leaves, and bond, polygon, rigid, and variation point modules are the interior nodes, see Fig.4. A process module which has no parent is a *root* of the C-MPSG.

Definition 3 (The C-MPSG). Let $P = (V, E, type)$ be a process graph. A tuple $H = (M, pl\_mod, mtype)$ is the Configurable Module-based Process Structure Graph, where: M is a set of process modules of P, such that Γ^M is a set of placeholders of process modules in M; $pl\_mod : \Gamma^M \rightarrow \mathcal{P}(M) \backslash \{\emptyset\}$ is a function which assigns to a placeholder of process module process modules in M; $mtype : M \rightarrow \{trivial, polygon, bond, rigid, var\_point\}$ is a function which assigns a type to a process module.

We say that the C-MPSG is structurally correct, if: (i) it has only one root (ii) polygon, bond, and rigid modules are in AND-relations with their child modules, and variation point modules are in XOR- and OR-relations with their child modules (iii) there is no cycles in the C-MPSG, otherwise the instantiation of placeholders of process modules will never terminate.

For any C-MPSG $H = (M, pl\_mod, mtype)$ a subset of process modules $V \subseteq M$ is selected out of the set of its process modules. The C-MPSG which corresponds to a subset of process modules of the C-MPSG is called the *variant C-MPSG* of a given C-MPSG, and the subset of process modules is called a *variant*. A variant C-MPSG has to be structurally correct. It needs to have the same root as a given C-MPSG. For any polygon, bond or rigid modules which is in the variant C-MPSG all its child process modules must be also in the variant C-MPSG. For any variation point module which is in XOR-relation with its child modules only one of its child process modules must be in the variant C-MPSG. For any variation point module which is in OR-relation with its child modules at least two of child process modules must be in the variant C-MPSG.

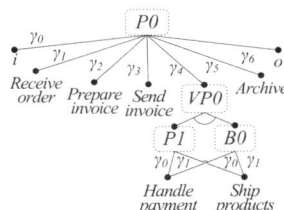

Fig. 4. The C-MPSG which captures two process variants from Fig.1

We introduce the individualization function $indv$ which maps the C-MPSG H and one of its variant V to the variant C-MPSG H^V, and the instantiation function $inst$ which maps the variant C-MPSG to a process graph. The semantic domain of the C-MPSG H with a set of available variants AV is a *family of the process variants* $\bigcup_{V \in AV} inst(indv(H, V))$.

An individual process variant can be specified by a user by selecting the relevant process modules from the integrated C-MPSG. To reduce the complexity of the C-MPSG and to provide an additional information to support the user decision between alternative process modules, we introduce an *annotated configuration view* which represents only variation point modules with their alternative child modules annotated with the rationale behind (see Fig.5). After all alternatives were assigned to variation point modules, the variant C-MPSG is derived out of integrated C-MPSG by applying the *individualization* operation. As the last

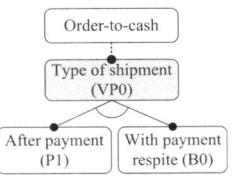

Fig. 5. Annotated configuration view for the C-MPSG from Fig.4

steps of configuration a desired process variant is reconstructed from the variant C-MPSG by applying the *instantiation* operation.

4.2   Design and Customization of the C-MPSG

The high-level steps which are performed for construction of the C-MPSG are presented in Fig.6. For a given set of process variants the first step of construction is to transform each process variant into the C-MPSG. We use the *modularization* operation which first constructs the RPST of a given process variant and then transforms obtained RPST into the individual C-MPSG. As the next step the constructed C-MPSGs have to be merged into the integrated C-MPSG. In order to avoid duplication and to increase reuse the common process modules have to be fused during the merge of the C-MPSGs. Thus, as a merge prerequisite, a *set of correspondences between process modules* has to be defined. To detect correspondences we pairwise compare process modules which have the same type, and for polygon modules we compare their subsequences. To compare two process modules we first evaluate the similarity degree between them by calculating a *similarity measure*, and next we check their conformance to a set of *correspondence rules*. As the result of comparison a set of correspondences between process modules is created. During the merge process modules which correspond to each other are fused in one module. For each point in the integrated C-MPSG where variation occurs a variation point module is created. After the merge, the set of available variants is generated.

Customization of the C-MPSG includes addition of a new process variant to the family of the process variants or removal of existing one. A new process variant can be create in two ways: as a recombination of existing process modules, or by arbitrary change of the the existing process variant. In the first case, a new process variant will be added to the C-MPSG by adding a new combination of process modules into a set of available variant of the C-MPSG. In the second case a process variant has to be modularized, compared, and merged with the integrated C-MPSG by applying the set of operations used during construction of the integrated C-MPSG.

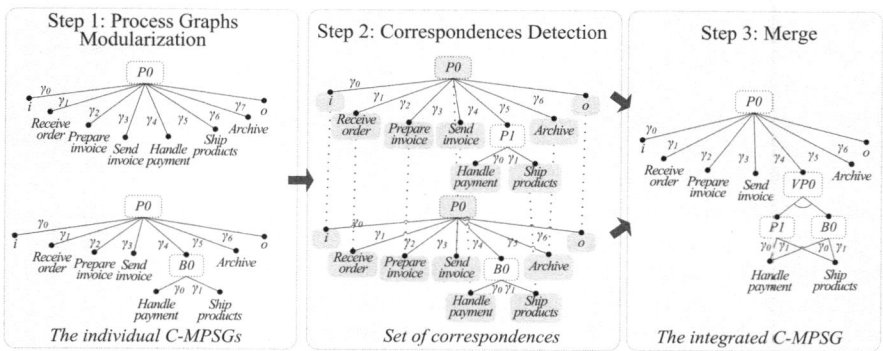

Fig. 6. Main steps which are performed for construction of the integrated C-MPSG

5   Conclusion and Future Work

In this work we have presented an approach for managing variability in process models. We proposed to decompose process variants into a set of reusable blocks, called process modules, and to capture a multiple process variants as an integrated set of process modules. In particular, in Sec.4.1 we gave a formal definition to process modules, the C-MPSG syntax and semantics, and described how the C-MPSG can be configured to a specific solution, while in Sec.4.2 we presented an overview on how the C-MPSG which integrates a family of process variant is constructed and further customized. The presented approach, due to its simplified process formalism can be adopted to various process modeling languages (e.g., BPMN [8], EPC [7]). The next step of our work is a prototypical implementation of the presented approach and validation of its applicability in practice.

References

1. Becker, J., Delfmann, P., Knackstedt, R.: Adaptive Reference Modeling: Integrating Configurative and Generic Adaptation Techniques for Information Models. In: RM 2006, pp. 27–58. Springer, Heidelberg (2006)
2. Gottschalk, F., van der Aalst, W.M.P., Jansen-Vullers, M.H., La Rosa, M.: Configurable Workflow Models. International Journal of Cooperative Information Systems 17(2), 177–221 (2008)
3. Gottschalk, F., van der Aalst, W.M.P., Jansen-Vullers, M.H.: Merging Event-driven Process Chains. In: Meersman, R., Tari, Z. (eds.) OTM 2008, Part I. LNCS, vol. 5331, pp. 418–426. Springer, Heidelberg (2008)
4. Gottschalk, F., Wagemakers, T.A.C., Jansen-Vullers, M.H., van der Aalst, W.M.P., La Rosa, M.: Configurable Process Models – Experiences from a Municipality Case Study. In: van Eck, P., Gordijn, J., Wieringa, R. (eds.) CAiSE 2009. LNCS, vol. 5565, pp. 486–500. Springer, Heidelberg (2009)
5. Jansen-Vullers, M.H., van der Aalst, W.M.P., Rosemann, M.: Mining configurable enterprise information systems. DKE 56(3), 195–244 (2006)
6. Hallerbach, A., Bauer, T., Reichert, M.: Managing process variants in the process lifecycle. In: Proc. 10th Int. Conf. on Enterprise Information Systems, pp. 154–161 (2008)
7. Mendling, J.: Metrics for Process Models: Empirical Foundations of Verification. In: Error Prediction and Guidelines for Correctness. LNBIP, vol. 6. Springer, Heidelberg (2008)
8. Object Management Group. Business Process Modeling Notation (BPMN), Version 1.1. OMG Specification. OMG (2008)
9. Polyvyanyy, A., Vanhatalo, J., Voelzer, H.: Simplified computation and generalization of the refined process structure tree. Technical Report RZ 3745, IBM (2009)
10. Rosemann, M., van der Aalst, W.M.P.: A configurable reference modelling language. Information Systems 32(1), 1–23 (2007)
11. Schnieders, A., Puhlmann, F.: Variability Mechanisms in E-Business Process Families. In: Proc. 9th Int. Conf. on Business Information Systems, pp. 583–601 (2006)
12. Vanhatalo, J., Voelzer, H., Koehler, J.: The refined process structure tree. In: Dumas, M., Reichert, M., Shan, M.-C. (eds.) BPM 2008. LNCS, vol. 5240, pp. 100–115. Springer, Heidelberg (2008)
13. Weske, M.: Business Process Management: Concepts, Languages, Architectures. Springer, Heidelberg (2007)

Adapting BPMN to Public Administration[*]

Victoria Torres, Pau Giner, Begoña Bonet, and Vicente Pelechano

Centro de Investigación en Métodos de Producción de Software,
Universidad Politécnica de Valencia,
46022 Valencia, Spain
{vtorres,pginer,pele}@pros.upv.es
Conselleria de Infraestructuras y Transporte,
Generalitat Valenciana,
46010 Valencia, Spain
bonet_beg@gva.es

Abstract. BPMN was originally defined to provide a process modelling notation that could be easily understood by all business stakeholders. However, when using this notation within a particular domain, as is the case with public administration, some limitations in the original notation become apparent. Documents play a very important role in the public domain. They become first-class citizens and standard BPMN notation is not sufficient to address certain common situations. This paper presents the extension defined for the BPMN 1.1 notation which is to be used to define process models in the Valencian public administration in Spain. In addition to this extension, the paper also presents how BPMN has been integrated into a software development methodology.

Keywords: BPMN, Public Administration, Document-centric Process Models.

1 Introduction

This paper presents the joint work that is currently being carried out by the Valencian Regional Ministry of Infrastructure and Transport (Conselleria de Infraestructuras y Transporte, hereafter CIT) and the *Centro de Investigación en Métodos de Producción de Software* (ProS research centre) to integrate the BPMN notation into the gvMétrica [1] methodology (based on METRICA III [2]). The CIT is developing the Modelling Software Kit (MOSKitt) [3] tool, which is an open source CASE tool built on the Eclipse [4] platform, to give support to the gvMétrica methodology. According to this methodology, MOSKitt supports the definition of different model types which include business process models. The STP Eclipse project [5] provides the foundations for the business process modelling tools available in MOSKitt. Most

---

[*] This work has been carried out with the support of the Spanish Ministry of Science and Innovation (MICINN) as part of the SESAMO TIN2007-62894 project. It has been co-financed by the ERDF, and by the Valencian Regional Ministry of Infrastructure and Transport through the Valencian Region's Operational Programme (*Programa Operativo de la Comunitat Valenciana*) 2007-2013.

J. Mendling, M. Weidlich, and M. Weske (Eds.): BPMN 2010, LNBIP 67, pp. 114–120, 2010.

of the process models built at the CIT capture the procedures that are associated with the services offered to its citizens. An important particularity of these processes is their document-oriented nature. In these processes, documents go back and forth between different CIT departments and the citizens involved in these processes. In addition, multiple copies of original documents are delivered to different process participants throughout the process. Keeping the flow of these documents under control becomes a must in Public Administration processes. To make all these requirements explicit, we hereby present the extension for the BPMN 1.1 notation [6] and the tool support that has been defined to satisfy CIT requirements.

The remainder of the paper has been structured as follows. Section 2 provides an overview of the gvMétrica methodology centred mainly on business process modelling and sets out the requirements that were put forward by the CIT for the notation and tool used to define business processes. Section 3 presents the solutions implemented in the MOSKitt BPMN editor to satisfy these requirements. Finally, section 4 summarises the work presented and draws some conclusions about its development.

2   Requirements for BP Modelling at the CIT

The gvMétrica methodology proposes the participation of different types of users during the software development process. More specifically, the business process modelling stage involves two different types of users: business analysts (personnel from the Organisation Department) and computer science engineers (personnel from the IT Department). On one hand, business analysts have thorough knowledge of the business and they need a language/notation and a supporting tool that allows them to represent this knowledge. On the other hand, computer science engineers are less interested in process semantics and are more focused on their correct performance in terms of a specific process execution engine. In addition to the business modelling task, the gvMétrica methodology proposes the construction of a set of different models which are intended to explicitly represent *roles* and their structure within the *organisation* and *documents*. These models increase the expressiveness of the business process model built in BPMN.

This section presents the requirements that business analysts from the CIT's Organisation Department have requested to correctly model their business processes. These requirements are presented first by giving a brief definition and then by providing the rationale behind them. These requirements refer not only to the notation itself but also to the corresponding editing support.

Requirement 1 (Augmenting semantics of documents involved in process models). It is necessary to link documents used in a process model with documents defined in the *Document model*. This linkage allows stakeholders to easily reach the referenced objects (and all their associated properties) that exist in the *Document model*.

Rationale. Documents play a very important role in CIT procedures. In fact, they conserve the data that is interchanged between the CIT and the related external participants. Therefore, tools are required to ensure easy access and handling of the documents that are involved in the process.

Requirement 2 (Dealing with multiple copies). It is essential to differentiate between original documents and copies and to provide modelling mechanisms that specify *when* copies can be distributed and to *whom* they should be sent.

Rationale. Most of the procedures defined in public administration usually involve producing different copies of the same document that have to be delivered to different departments. Therefore, it is very important to keep track of all these copies and to ensure that they reach their correct destination.

Requirement 3 (Dealing with document status). It is necessary to provide modelling mechanisms that allow the current status of a document involved in the process model to be specified.

Rationale. Every document received at the CIT and involved in any of its processes is given a status. The value of this status changes across the life-cycle of the document during the process. The possible status values: *submitted, issued, signed* and *archived,* represent the document life-cycle.

Requirement 4 (Grouping documents). It is necessary to provide modelling mechanisms that allow groups of documents that flow as a unit through the process to be specified.

Rationale. In addition to single documents, set of documents (documents which only make sense as part of a specific set) commonly flow through the process as a unit. This is why the introduction of the group of documents concept to refer to this inseparable set, in addition to the document concept, is essential.

Requirement 5 (Specifying task semantics). The semantics of process tasks need to be explicitly defined not only for visual purposes but also for their use during the handling of the process model when applying the methodology.

Rationale. In the CIT, process models built during gvMétrica projects evolve with the different type of stakeholders involved in the modelling process and because they must finally be executed in a process engine. Therefore, providing specialisation of intent of the model elements enables enhanced model comprehension and improves subsequent handling.

Requirement 6 (Augmenting semantics of roles involved in process models). Tool mechanisms need to be provided to link the participants defined in a business process model with the hierarchical structure of the organisation.

Rationale. Governments have a well-defined organisational structure. This structure can include inheritance relationships between different positions in the organisation (e.g. technician sub-types) thus enabling the use of abstract (general technician) and specific (IT technician) roles in a consistent manner. Therefore, associating process modelling elements (e.g. pools and lanes) with this structure lends more semantics to these elements and to the process itself.

Requirement 7 (Flow integration with external participants). It is necessary to provide modelling mechanisms that enable the control flow between the tasks performed by external and internal participants involved in the process to be defined.

Rationale. In most of the procedures modelled at the CIT, the tasks performed by external participants (citizens, outsourcing companies, other hierarchical departments within the CIT) are involved in the process's control flow. However, if we represent external participants as a separate pool, this involvement cannot be expressed considering the expressiveness provided by the BPMN notation.

Requirement 8 (Dealing with the status of a process model). Modelling mechanisms need to be provided that enable the life-cycle of the process model definition to be specified.

Rationale. From their conception to their use process models go through various stages e.g. *draft, in-study, definitive, obsolete* (deprecated by users) and *expired* (deprecated by law).

3 Solutions to Support CIT Requirements

In line with the structure of the previous section, this section puts forward the solutions that have been defined and that are being implemented (some are already available in current versions of MOSKitt) to satisfy the aforementioned requirements.

Most of the solutions consist of defining extensions for the original BPMN notation. These extensions have been implemented following a non-invasive approach (by using the *Eclipse Extension Points* and *Extensions* provided in Eclipse). That is, the original meta-model (the one included in the STP project) has not been modified. Therefore, the BPMN models built in MOSKitt can still be handled by editors implementing the original notation. Figure 1 illustrates some of the requirements presented in this section.

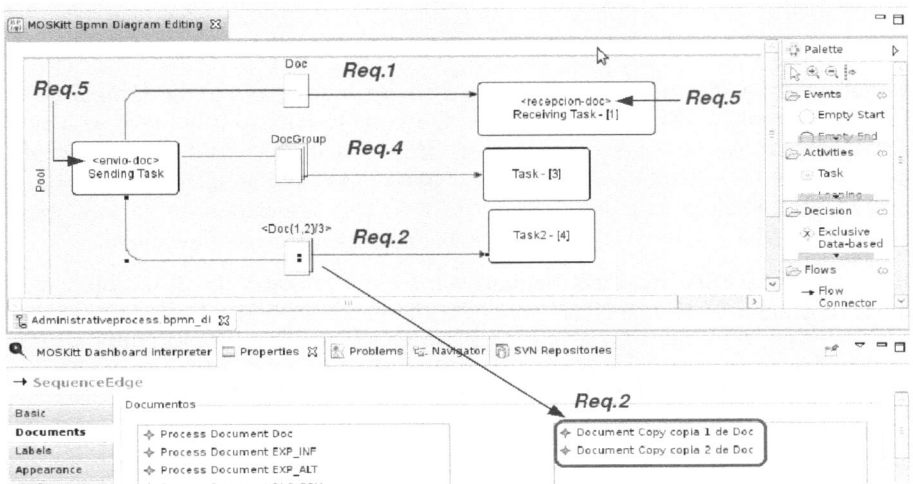

Fig. 1. Data objects used in the BPMN MOSKitt editor

Requirement 1 (Augmenting semantics of documents involved in process models).
BPMN provides the *Data Object* element to represent documents, data and other
objects that are handled during the process. The semantics of such elements can be
provided by the modeller and in the case of the CIT, this type of elements has been
extended in two ways. On one hand, a new property has been defined to augment
document semantics. This property enables the graphic element to be associated with
a document defined in the *Document model*. On the other hand, its graphic
representation has been slightly changed to reduce the complexity of the resulting
models (see document representation in Figure 1). A new modelling element (called
Information flow) has been introduced into the editor palette. This element is
semantically equivalent to merging a *Sequence flow* element and a *Data Object*
element and reduces complexity since the number of lines connecting elements is
drastically reduced.

Requirement 2 (Dealing with multiple copies). As stated in requirement 1, within
the context of the CIT, *Data Objects* always refer to documents that are used
throughout the process. However, within the process model a difference must be
made between original documents and their copies. In MOSKitt, to explicitly define
that a document refers to a specific copy, the document is labelled according to a
specific pattern (see copy representation in Figure 1). By making this differentiation
graphically explicit, users find it easier to identify document copies which are flowing
in the process.

Requirement 3 (Dealing with document status). In BPMN, *Data Objects* add to the
common *Artifact* attributes with three new attributes which include the *status*
attribute. This attribute was conceived to indicate the impact the process has on the
Data Object. However, in the context of the CIT, the valid status in which documents
may exist are clear. Therefore, the editor provides a set of predefined values which
are associated with this attribute. The set of possible values of this attribute
corresponds to *received, issued, stamped* and *archived*. These are the valid stages in
which documents handled in a process may exist.

Requirement 4 (Grouping documents). The flexibility given to the *Data Object* in
BPMN to represent any kind of resource allows us to refer to it not only as a single
document but also as a set of documents. In MOSKitt, sets of documents are first
defined in the *Document model*. Then, *Information flow* elements defined in the
process model can be associated with these sets. This association results in a graphic
change (see Req.4 in Figure 1) which enhances comprehension of the model.

Requirement 5 (Specify Task Semantics). BPMN provides the *Task* object as the
modelling element to represent atomic work that is performed within a business
process. This type of objects can be categorised into one of the different types
identified in the specification, which are *Service, Receive, Send, User, Script, Manual,
Reference* and *None*. However, in real-case scenarios we found that there is a need for
a trade-off in the definition of task types. On one hand, not all standard BPMN types
are required by the organisation and providing editing and execution support for them
only increases the complexity of the system. On the other hand, some activity types
need to be specialised to cater for particular organisational needs. This not only
increases the precision of the diagrams but also enables specific execution support.

For example, the *send* type has been divided into several types which differentiate what type of element is being sent (see task decoration in Figure 1). Therefore, to indicate that a specific task only involves the sending of a document, the task is typified as *send-doc*. When referring to tasks that only involve receiving a document, the task is typified as *receive-doc*.

Requirement 6 (Augmenting semantics of roles involved in process models). In this case, the editor provides a linkage between the BPMN model and the *Role model*, along the same lines as the linkage established between the BPMN model and the *Document model*. This linkage is possible because we have defined a new property which is associated with the lane element that enables us to indicate which *role* in the organisational structure is responsible for this element. As a result of this linkage, the lane element of the process diagram is automatically updated by showing the name of the selected role in its associated label.

Requirement 7 (Flow integration with external participants). The need to involve external participants in the internal process flow forced us to model external participants as a lane within the organisation. However, we have added a new property to the lane element to differentiate them from internal roles. This property is named *external* and when it is selected, it is automatically decorated with the <<external>> stereotype. This decoration allows us to graphically differentiate the types of roles participating in a process. Distinguishing between external and internal organisations was a well-known requirement for executability in BPMN process execution engines such as Intalio BPMS.

Requirement 8 (Dealing with the status of a process model). To specify the current status of a process we have added a new attribute to the process element. This is called status and the available values correspond to the different status that constitute the process life-cycle which are: *draft*, *in-study*, *definitive*, *obsolete* and *expired*.

4 Conclusions

This paper has presented the extension defined for the BPMN 1.1 notation to satisfy the requirements put forward by the CIT to define their process models. The major feature of these models is their document-centric nature. This is why new modelling mechanisms had to be added to the notation to provide all the required expressiveness. The latest released specification of BPMN [8], which is still at a beta stage, takes into account the fact that process models are normally not considered in isolation but as part of a more complex system development process. Thus, this new version of the standard provides a new element that enables *BPMN Artifacts* to be integrated into a development process. This new element, called *External Relationship*, allows *BPMN Artifacts* to be associated with elements from any other domain model. In line with this argument, in gvMétrica we faced this necessity and solved it by means of the facilities provided by the Eclipse platform, that is, by means of *Extension Points* and *Extensions*. In addition to this extension, new functionalities have been implemented to improve the usability of the BPMN editor. The goals of these improvements are (1) to speed up the construction of BPMN models, (2) to augment the semantics of the BPMN elements by linking BPMN process models with other models defined by gvMétrica, and (3) to improve the comprehension of the model by decorating the BPMN diagrams. Currently, the MOSKitt tool is being used in an industrial environment where the user's feedback is continuously taken into account to improve it. Even though the MOSKitt project has been developed within the CIT context, the members involved in it have thorough knowledge of Valencian Public Administration. Therefore, the requirements presented in this work are not CIT-specific but can also be applied to other Valencian regional ministries.

References

1. gvMétrica,
   http://www.gvpontis.gva.es/proyectos-integra/
   metodologias-aplicadas/gvmetrica-adaptacion/
2. Métrica III, http://www.csi.map.es/csi/metrica3/index.html
3. MOSKitt. Modeling Software Kitt, http://www.moskitt.org/
4. Eclipse Software Development Environment, http://www.eclipse.org/
5. STP BPMN Modelder, http://www.eclipse.org/bpmn/
6. Business Process Modeling Notation, V1.1. OMG Document Number: formal/2008-01-17
7. zur Muehlen, M., Recker, J.: How much language is enough? theoretical and practical use of the business process modeling notation. In: Bellahsène, Z., Léonard, M. (eds.) CAiSE 2008. LNCS, vol. 5074, pp. 465–479. Springer, Heidelberg (2008)
8. Business Process Model and Notation, V2.0. OMG Document Number: dtc/2009-08-14

An Evaluation of BPMN Modeling Tools

Zhiqiang Yan, Hajo A. Reijers, and Remco M. Dijkman

Eindhoven University of Technology
PO Box 513, NL-5600 MB Eindhoven, The Netherlands
{z.yan,h.a.reijers,r.m.dijkman}@tue.nl

Abstract. Various BPMN modeling tools are available and it is close to impossible to understand their functional differences without simply trying them out. This paper presents an evaluation framework and presents the outcomes of its application to a set of five BPMN modeling tools. We report on various differences, for example with respect to the level of BPMN support and the "smart" support a modeler may experience using these tools. Interestingly, while tools that are provided by commercial parties do offer more integration options with other tools in BPM suites, they cannot be said to outperform academic tools in other respects.

1 Introduction

Along with the rising popularity of the BPMN standard, the number of tools increases that support the modeling of process models in this notation. The existence of a standard, however, does not fully determine the functionality of tools claiming to support that standard. Different interpretations of a standard may generate different implementations – the behaviorial differences of BPEL engines is a painful case in point [4]. But even beyond interpretation issues, developers of a modeling tool will have considerable freedom with respect to the *coverage* of the standard (e.g. core vs. a subset), *user interface* design (e.g. support to place elements on the canvas), *model management* features (e.g. storage location), *licensing policies* (e.g. freeware vs. commercial rates), and other aspects.

The evaluation consists of the following steps. First, five popular BPMN modeling tools are selected, including two academic and three commercial tools. A specific consideration from our side for the inclusion of tools in this evaluation is that we were interested in potential differences between tools provided by academic and commercial parties. Second, an evaluation framework is presented, which is derived from the framework of Nüttgens [8]. The focus of our evaluation is on the *functional* aspects of BPMN modeling tools. As a result, our evaluation does not cover performance-relates issues (e.g. retrieval speed, repository size, etc.). Last, selected tools are evaluated based on the framework. The evaluation results are obtained by both checking the documentations of the tools and modeling processes using the tools ourselves.

By presenting a contemporary evaluation of a set of popular BPMN modeling tools, the purpose with this paper is to see the functional differences of these tools, especially the functional differences between academic and commercial

J. Mendling, M. Weidlich, and M. Weske (Eds.): BPMN 2010, LNBIP 67, pp. 121–128, 2010.

tools. Since we are not aware of any comparable efforts, the evaluation outcomes may be useful input for modelers that are currently looking for a tool
that supports them in developing and editing BPMN diagrams[1]. Furthermore,
researchers that are interested in further or more thorough evaluations of BPMN
tools may find inspiration in the presented evaluation framework.

The structure of this paper is as follows. In Section 2, we present and motivate
an evaluation framework for BPMN modeling tools. This framework is applied
to five modeling tools, as is described in Section 3. Section 4 concludes.

2   BPMN Modeling Tool Evaluation Framework

A framework for evaluating BPMN modeling tools is proposed in this section.
The framework identifies a number of aspects that are considered relevant for
the evaluation and criteria that determine the rules to evaluate these tools with
respect to every aspect.

2.1   Evaluation Aspects

The aspects that are considered relevant for the evaluation of BPMN modeling tools are derived from the evaluation components proposed by Nüttgens for
business process modeling tools [8]. Nüttgens' framework is designed for general business process modeling, which makes it applicable to the BPMN modeling evaluation with some necessary adaptations, e.g., evaluating BPMN support
(version and elements). There are five components in Nüttgens' framework: product and pricing model, producer and customer base, technology and interfaces,
methodology and modeling, and application and integration. The focus of this
paper is on the functional aspects of BPMN modeling tools. Consequently, we
focus on the last two components of Nüttgens' framework: methodology and
modeling, and application and integration. We use the elements of these components as aspects in our framework and we refer to these elements as function
aspects. The first three components of Nüttgens' framework concern product
information, which is not the focus of this paper. Therefore, we combine these
three components and select a few essential elements, e.g. licence type, as the
product aspects for the framework in this paper. As a result, the aspects that
are considered relevant for the evaluation of BPMN modeling tools in this paper
are the following.

- Product Aspects
  - Main Users; Licence; Localization
- Function Aspects
  - Modeling Support
    * BPMN Version; BPMN Elements; Modeling Notation
  - Model Creation

---

[1] Note that we assume that such modelers are not satisfied with simply using MS
Visio with a BPMN stencil, as indeed they should not be.

  * Element Groups; Reusable Patterns; Quick Completion
- Model Navigation
  * Automatic Layout; Zoom-in/out; Birds-eye; Layout Direction
- Model Storage
  * Web-based; Repository; Import/Export Format
- Integration

As relevant product aspects we identify the main type of intended users of the tools, the type of license of the tool, and the localizations (e.g. natural languages, decimal system) that are supported by the tools. As relevant functional aspects we identify the support that the tool has for: BPMN and possibly other modeling languages; easy creation of models; navigation of models that are created; storage of models; and the integration with other tools. These aspects are further divided.

For the support that a tool has for BPMN and other modeling languages we identify: the version of BPMN that the tool supports, the (subset of) BPMN elements of that version that the tool supports and possible other languages that the tool supports. With respect to the ease with which models can be created with the tool, we investigate whether the tool has support for organizing model elements into groups (e.g.: stencils in Visio), whether the tool contains re-usable model fragments or patterns and whether the tool supports quick completion of models. Quick completion allows for the creation of multiple elements at once, typically a control flow relation and a certain control flow node at the same time. For the support that a tool has for model navigation, we determine whether or not the tool supports automatic layout, zooming into and out of a model, a birds-eye view of the model and the layout directions that a tool supports. For model storage we investigate: whether the modeling 'client' can run in a web-browser or needs to be installed on a machine on the client-side; whether the tool has a local repository or a repository that can be shared by multiple users and what import and export formats the tool supports.

2.2 Evaluation Criteria

Each aspect for evaluation can have different values. In this section we present the different values and the criteria that a tool must meet to have a certain value. The possible values per aspect are shown in Table 1. To some extent the possible values may be open, in which case we only enumerate the most important cases that we observed in the tools that we studied. For example, integration with many different types of tools can be envisioned. However, for the tools that we studied we observed integration with (formal) verification tools and with BPM suites, which is why these two values are listed.

We distinguish between tools that focus on academic users and industrial users, where we consider tools that focus on academic users to be research prototypes that focus more on advanced functionality, while tools that focus on industrial users have a stronger focus on proven functionality and on stability and performance. With respect to types of licenses, we distinguish between commercial (i.e. paid), free and open source licences. For the localization aspect, the tool can support only English-language localizations or more localizations, e.g., Spanish and Chinese. The BPMN versions can be supported by a tool are 1.0, 1.1, 1.2, and 2.0. A tool can either support all elements of a specific BPMN version or a subset of those elements. We distinguish between the simple, standard and complete class of elements [11]. The simple class contains tasks, subprocesses, exclusive data-based and parallel gateways, start and end event without a specific trigger or result, pools and lanes, artifacts and sequence-flow. In addition to the elements from the simple class, the standard class contains typing and decoration of tasks and subprocesses, the inclusive and the event-based gateway, intermediate events and timer, message and error triggers and message flows. The complete class contains all BPMN elements. Of course all tools that we evaluate support the BPMN notation, but some tools also support other notations. The most supported notations by process modeling tools [10] are used as the criteria: BPMN, EPC and Petri nets. There are different forms of assistance for creating BPMN models, which can either be provided by a tool or not. There are also different forms of assistance for navigating a BPMN model that can either be provided or not and the (automatic) layout direction can either be left-right, top-down or both. With respect to the storage of models we distinguish between tools that have a web-based client (i.e. can be run in a browser) and tools that do not and for which a client tool must be installed. We also distinguish between tools that have a shared repository and tools for which the model repository exists locally on the client machine. The tools can support several formats for import and export of models from their repository, of which XML, BPEL, XPDL are the most popular ones [10]. We observed integration of the BPMN modeling tools with both model verification tools and complete BPM suites that provide comprehensive support for the complete business process lifecycle, from modeling to enactment, monitoring and re-engineering.

3   BPMN Modeling Tool Evaluation

We evaluated five BPMN modeling tools for this paper. To the best of our knowledge, Oryx [3,9] and DiaGen [2,7] are the major academic modeling tools that focus on BPMN. Biz-Agi [1], Tibco [12,13] and Intalio [5,6] are three popular industrial BPMN modeling tools (e.g., over 50,000 people have downloaded Tibco [13]). We performed the evaluation in two steps: first we got information from the websites, forums, documentations and the tools themselves to compare them based on the evaluation framework proposed in Section 2. Second, we created BPMN models with these tools to get some concrete experience and to be able to discuss the user experience of working with these tools.

The evaluation results based on the framework are shown in Table 2. The rows represent the evaluation aspects and the columns represent the tools evaluated in this paper. The users of Oryx and DiaGen are mainly from the academic field (although Oryx has a commercial spin-off) and the main users of the other three tools are from the industry. All the tools provide a free version; Oryx, DiaGen and a version of Intalio are open-source; Intalio also provides a professional version

Table 1. BPMN Modeling Tool Evaluation Criteria

Product Aspects			
Main Users	Academic	Industrial	
Licence	Commercial	Free	Open source
Localization	English	More	
Function Aspects			
BPMN Version	v1.0	v1.1	v1.2	v2.0
BPMN Elements	Simple	Standard	
Modeling Notation	BPMN	EPC	Petri nets
Element Groups	Not provided	Provided	
Reusable Patterns	Not provided	Provided	
Quick Completion	Not provided	Provided	
Automatic Layout	Not provided	Provided	
Zoom-in/out	Not provided	Provided	
Birds-eye	Not provided	Provided	
Layout Direction	Left-right	Top-down	
Web-based	Yes	No	
Repository	Local	Shared	
Format	XML	XPDL	BPEL
Integration	Verification Tool BPM Suite		

that users need to pay for. Most of the tools only support English, except for Biz-Agi, which supports many localizations.

Biz-Agi supports BPMN v1.1; DiaGen, Tibco and Intalio support BPMN v1.2; Oryx supports multiple versions of BPMN through different 'stencils': v1.0, v1.2 and the latest version of BPMN, BPMN v2.0. Oryx, Biz-Agi and Tibco support all BPMN modeling elements for their version. Intalio is fairly complete in the elements that it supports. However, interestingly it does not support boundary intermediate events on a task, while it does support boundary intermediate events on a subprocess. In addition to that it does not support properties that indicate the implementation type of a task (although different implementations of a task *are* supported). These restrictions are deliberate design choices of Intalio. However, it does limit the conformance of the tool to the BPMN standard. DiaGen supports a subset of the simple class of BPMN modeling elements. In particular it does not support subprocesses or artifacts. Oryx is the only tool that supports modeling notations besides BPMN. These notations include different variants of the EPC and Petri net modeling notations. In addition to that Oryx supports notations from other perspectives, such as data (UML class diagrams) and forms (XForms). However, these are out of the scope of this paper.

With respect to model creation, all of the tools divide the BPMN elements into several smaller, more manageable, groups except for DiaGen. Tibco contains

Table 2. BPMN Modeling Tool Evaluation

	Oryx	DiaGen	Biz-Agi	Tibco	Intalio
Product Aspects					
Main Users	Academic	Academic	Industrial	Industrial	Industrial
License	Free,Open	Free,Open	Free	Free	*All*
Localization	English	English	Multiple	English	English
Function Aspects					
BPMN Version	v1.0,v1.2,v2.0	v1.2	v1.1	v1.2	v1.2
BPMN Elements	Complete	Simple-	Complete	Complete	Standard-
Modeling Notation	*All*	BPMN	BPMN	BPMN	BPMN
Element Groups	Provided	Not provided	Provided	Provided	Provided
Reusable Patterns	Not provided	Not provided	Not provided	Provided	Not provided
Quick Completion	Provided	Provided	Provided	Not provided	Not provided
Automatic Layout	Provided	Provided	Provided	Provided	Provided
Zoom-in/out	Provided	Provided	Provided	Provided	Provided
Layout Direction	Left-right	Left-right	Left-right	Left-right	Left-right
Web-based	Yes	No	No	No	No
Repository	Local,Shared	Local	Local	Local	Local,Shared
Format	XML,XPDL	XML	XML,XPDL	*All*	XML,BPEL
Integration	Verification	Verification	BPM Suite	BPM Suite	BPM Suite

("*All*" represents "all the options listed in Table 1", "-" represents "less than")

a library of small model fragments that can be re-used when modeling. DiaGen, Oryx and Biz-Agi provide support for quick completion of models.

All tools provide automatic layout. Also, all tools can zoom in/out and the layout direction for all tools is from left to right.

Oryx is the only tool for which the client can be accessed through a web browser. The other tools require installation. All of the tools support storing BPMN models locally on the client's machine, while Oryx and Intalio also provide a shared repository. For import and export of models all of the tools support some XML-based format; Oryx, Biz-Agi and Tibco support XPDL. Tibco and Intalio support BPEL. Some of the tools also support other formats that are not listed in the table, e.g., Oryx can also import from ERDF and JSON formats, and export to ERDF, JSON, RDF, PNML, XMI and PDF formats.

With respect to integration with other tools, Biz-Agi, Tibco and Intalio integrate with their own BPM suites, and therefore provide integration with tools from the complete BPM lifecycle. Oryx and DiaGen mainly focus on modeling, only providing simple integration for process mode verification.

Two BPMN models are selected to evaluate the user experience of working with the tools, which are used to check whether a tool can support the simple class and the standard class of BPMN elements in [11]. These two models differ in size and complexity, e.g., the former one with 5 tasks, 2 events, 2 gateways and 11 flows, while the later one with 9 tasks, 10 events, 4 gateways and 29 flows. From our experience of working with the tools, we derived the following conclusions.

Oryx is the only web-based BPMN modeling tool in our evaluation. The web-based characteristic makes it convenient to create models without installing the tool, but at the same time makes it highly dependent on the web browser, e.g., Oryx works well with Firefox, but does not work with Internet Explorer.

The automatic layout functionality that the tools provide can be applied to a small set of fragments of a BPMN model only. Oryx provides automatic layout for the whole model. However, it does not respond when the model is complex, e.g., the second model of the evaluation. Intalio incorporates additional functionality for automatically aligning a task with tasks that are already in the model at the moment that a task is inserted, while the other tools require that users first insert the task in the model and then press a button to align the task.

Oryx and Biz-Agi provide some smart suggestions during modeling. When a user clicks on an element, it shows possible elements that can connect to it. This is convenient, because the user does not need to go to the panel and find the element. DiaGen provides a modeling tool with syntax-based assistance, which can help to automatically combine or complete process fragments, help generate new processes, and help preserve correctness of process models. However, the intelligence of DiaGen is not always useful for creating a model (users may be not interested in syntax-based assistance).

Tibco and Intalio provide professional BPMN modeling tools, which can cover the whole BPM lifecycle (e.g., enactment) for process models rather than the modeling procedure. The drawback of this is that it takes relatively more time for a novice user to become familiar with these tools, because of their complexity, e.g., the user needs to setup a project before he can start modeling. Therefore, Oryx and Biz-Agi are easy for beginners to start with and also more convenient for users who only want to model and do not require support for the complete BPM lifecycle.

4   Conclusion

This paper presents a framework for evaluating the BPMN modeling tools with respect to functional aspects. Five tools are evaluated using the framework: Oryx, DiaGen, Biz-Agi, Tibco and Intalio. Oryx and DiaGen are academic prototype tools, while Biz-Agi, Tibco and Intalio are industrial tools.

The framework focuses on the functional aspects of the tools. These include the level of support that the tool has for the BPMN notation, functionality to assist the modeler in creating the models, functionality to assist in navigation through the models, storage technology that is used and functionality for interchanging models with other tools. In addition to that, the framework addresses some aspects to characterize the tool as a product. These include licensing and localization.

Evaluation of the BPMN modeling tools shows that the level of support that the tools provide for the BPMN notation differs for the various tools. Oryx is currently the only tool that supports BPMN 2.0, the other tools support BPMN 1.1 or 1.2. DiaGen and Intalio support a subset of the BPMN notational elements, while the other tools offer complete support for their version of BPMN.

With respect to model navigation capabilities, all of the tools support automatic layout functionality and functionality for zooming.

With respect to storage functionality, Oryx differentiates itself from the rest by providing a web-based solution for modeling. Consequently, it also uses a central model repository that can be shared by multiple modelers, enabling collaboration. Intalio also offers a shared repository option. All of the tools support one or more open formats for interchanging models with other tools.

We expected that there would be a difference between the functionality offered by academic and industrial tools. However, that turned out not to be the case. Although there are differences between the level of BPMN support and storage functionality that is offered by tools, these differences do not depend on whether a tool is an academic or an industrial tool. The main difference between academic and industrial tools is that industrial tools are typically part of a larger BPM suite that provides support for process simulation, automation and monitoring, while academic tools focus solely on the modeling of processes.

References

1. BizAgi, http://www.bizagi.com/
2. DiaGen, http://www.unibw.de/inf2/DiaGen/assistance/bpm
3. Decker, G., Overdick, H., Weske, M.: Oryx C An Open Modeling Platform for the BPM Community. In: Dumas, M., Reichert, M., Shan, M.-C. (eds.) BPM 2008. LNCS, vol. 5240, pp. 382–385. Springer, Heidelberg (2008)
4. Hallwyl, T., Henglein, F., Hildebrandt, T.: A Standard-driven Implementation of WS-BPEL 2.0. In: Proceedings of SAC 2010, Sierre, Switzerland, pp. 2472–2476 (2010)
5. Helkiö, P., Seppälä, A., Syd, O.: Evaluation of Intalio BPM Tool, T-86.5161 Special Course in Information System Integration (2008)
6. Intalio, http://www.intalio.com/
7. Mazanek, S., Minas, M.: Business Process Models as a Showcase for Syntax-based Assistance in Diagram Editors. In: Schürr, A., Selic, B. (eds.) MODELS 2009. LNCS, vol. 5795, pp. 322–336. Springer, Heidelberg (2009)
8. Nüttgens, M.: An Evaluation Framework for Business Process Modelling Tools, http://www.cimosa.de/IctSupport/ToolSt-mn02.html
9. Oryx, http://bpt.hpi.uni-potsdam.de/Oryx/WebHome
10. Schmietendorf, A.: Assessment of Business Process Modeling Tools under Consideration of Business Process Management Activities. In: Dumke, R.R., Braungarten, R., Büren, G., Abran, A., Cuadrado-Gallego, J.J. (eds.) IWSM 2008. LNCS, vol. 5338, pp. 141–154. Springer, Heidelberg (2008)
11. Shapiro, R.: Model Portability Overview and Goals (2009)
12. Tibco Team. Business Process Modeling, http://www.tibco.com/multimedia/business-process-modelling_tcm8-2404.pdf
13. Tibco, http://www.tibco.com/products/bpm/process-modeling/default.jsp

Author Index

Bonet, Begoña   114

Decker, Gero   63
Dijkman, Remco M.   16, 121
Dumas, Marlon   1

Effinger, Philip   31

Fickinger, Tobias   78

García-Bañuelos, Luciano   1
Giner, Pau   114

Jogsch, Nicole   31

Koehler, Jana   46
Krumnow, Stefan   63
Kunz, Steffen   78

Leymann, Frank   8

Pascalau, Emilian   91
Pelechano, Vicente   114
Polyvyanyy, Artem   1
Prescher, Johannes   78

Rastrepkina, Maria   106
Rath, Clemens   91
Reijers, Hajo A.   121

Seiz, Sandra   31
Spengler, Klaus   78

Torres, Victoria   114

Van Gorp, Pieter   16
Völzer, Hagen   14

Yan, Zhiqiang   121